NATURAL DYEING WITH PLANTS

Other Schiffer Books on Related Subjects:

Dyeing Wool: 20 Techniques, Beginner to Advanced,
Karen Schellinger, ISBN 978-0-7643-3432-0

Earthen Pigments: Hand-Gathering & Using Natural Colors in Art,
Sandy Webster, ISBN 978-0-7643-4178-6

The Big Book of Flax: A Compendium of Facts, Art, Lore, Projects, and Song,
Christian and Johannes Zinzendorf, ISBN 978-0-7643-3715-4

Originally published as *Natürlich Färben mit Pflanzen* by Leopold Stocker Verlag, Graz © 2016 Leopold Stocker Verlag
Translated from the German by Omicron Language Solutions, LLC.

Library of Congress Control Number: 2017954825

Cover design by Danielle Farmer
Cover photo: Mona Lorenz, Gmunden.
Photo credits: Wisteria photo, p. 90, Gertraud Hölzl-Harb. All other images, unless indicated otherwise, provided by the authors.
Type set in Avenir Next/Caecilia LT Std

ISBN: 978-0-7643-5517-2
Printed in China

Published by Schiffer Publishing, Ltd.
4880 Lower Valley Road
Atglen, PA 19310
Phone: (610) 593-1777; Fax: (610) 593-2002
E-mail: Info@schifferbooks.com
Web: www.schifferbooks.com

FRANZISKA EBNER | ROMANA HASENÖHRL

NATURAL DYEING WITH PLANTS

GLORIOUS COLORS FROM ROOTS, LEAVES & FLOWERS

4880 Lower Valley Road • Atglen, PA 19310

CONTENTS

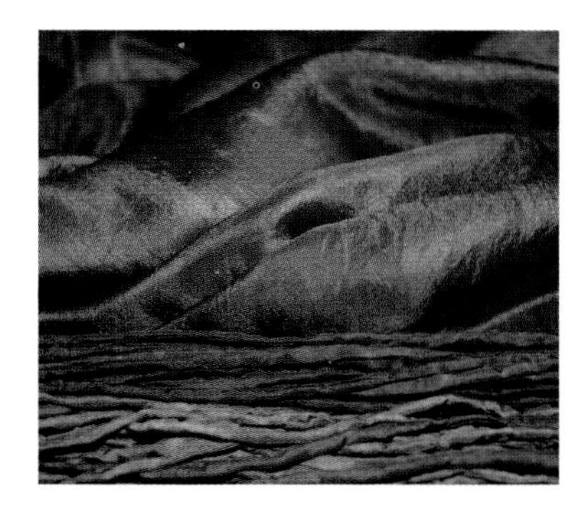

INTRODUCTION

Dyeing with plants is as old as the history of humankind. If we compare the time period during which humans have dyed their clothing using plants with the period of doing it using synthetic dyes, we find that dyeing with synthetics represents a tiny time period in this history. Why, then, have we gotten away from dyeworks using plants? Why do we prefer industrially dyed products to those that are made colorful with dyes from domestic plants?

There are many answers and some are obvious. The pace of human life is constantly increasing; time is short and time is precious. Dyeing with plant dyes, however, takes time and a loving approach to the product and the appropriate colors for it. In the twenty-first century, this seems to be possible only in a limited way, since the times have become too fast and the resulting demands are too great to use this technique effectively.

Another reason is the quantity of dyed goods and the associated financial factor. In times when dyeworks used plants, each person didn't own as many pieces of clothing

as today; clothes were of good quality and were patched, darned, and mended when necessary. This is hardly ever done today; the clothes we wear have a lifetime of one season and are so cheap that it isn't worth mending them. However, for clothing to be so cheap, it must be industrially manufactured and industrially dyed. And we can hardly avoid the fact that, at the end of the chain, there are people who work in bad conditions, since otherwise this cycle would be impossible.

It would also not be possible to dye the quantity of textiles and clothing we wear in the twenty-first century—we would quickly reach the limit of the dye resources. People's consumption behavior would have to change drastically, a development that isn't in sight. That is why this book should not be seen as a call to return completely to dyeing with plants. This book is intended to provide insight into the history and development of dyeing and the art of dyeing, and the possibilities we still have today for dyeing with plants.

This book provides an overview of what can be dyed with plants, how dyes are made, and how they dye different fabrics. Dyeing using plants is an art that has gradually re-emerged from the past and is again practiced intensively, particularly in the arts and crafts sector. Maybe this book will be an incentive for you to try this out yourself!

It's possible to create the entire color palette.

Glorious colors directly from nature.

HISTORY & TALES
Dyeworks in the Past

There are so many stories, sayings, legends, suppositions, and pieces of evidence concerning dyeing textiles with plant dyes. An art that has been practiced since the Stone Age brings all of these along with it, and beyond that, it shows us how deeply people have been involved in dyeing throughout human history.

It's rather difficult to determine just when exactly people began to dye textiles, because textiles do not last. It is much easier to reconstruct how dyes were used in cave paintings or for body painting. These dyeing activities were already being practiced by the time of the Neanderthals, so the use of colors for decoration, and the recording of stories on stone walls or for ritual purposes, is tens of thousands of years old!

Dyeing, painting, and drawing was verifiably done with yellow and red ocher, that is, red earth, which was used in many ways, including for rituals. The dyeing of textiles was clearly more difficult. We do not know whether the Neanderthal began attempts in this direction. Perhaps they painted leather? As with so many developments in the early days of mankind, we can no longer understand today how they came up with the idea of using dyes, and how many frustrating experiences were necessary until it finally worked. We only know that people had already started early on to dye their clothes and that they discovered how they had to handle the textiles, as well as the dye, to make this work.

What can be proven about the art of dyeing in antiquity? That achieves the leap through time to the present primarily in the form of grave offerings. Since Roman times, the dyeing of textiles has been very well documented in writing in the form of descriptions, recipes, and even contracts. What is clearly recognizable from the collected documentation, records, and pieces is that dyework is generally based upon **three groups of dyes**:

- *Dyes from minerals such as ocher*
- *Dyes from animals such as the scale insect and the purple snail*
- *Dyes from plants*

This book is concerned with the third group, that is, dyes from plants. The other two groups, together with plant dyeworks, represent an important part of the development of textile dyeworks. The dyes obtained from minerals are likely the oldest dyes used—as mentioned above, they go back to the time of the Neanderthals. Consistent documentation of dyeworks, however, has existed only since the time of the Romans. But what happened between the cave paintings and the Roman era? How and what did the Celts dye? And to what extent did the writings of ancient authors influence the dye recipes of the Middle Ages? In this book, we look into these questions, especially regarding Central Europe. Other cultures are considered if there is a demonstrable influence on the history of European dyeing culture.

Ancient peoples also may have used the sun to help with dyeing.

OF SETTLED HUNTERS AND VALIANT CELTS

The professional processing of textiles into clothing can only be established at the point when mankind became settled. This happened at different times in different places in the world. Here we are to scrutinize Central Europe more closely, but always in comparison to other parts of the world. There is indeed evidence in Central Europe that plant fibers such as grasses and basts were already being processed during the Paleolithic period, but it was not yet possible to manufacture complex fabrics at this time. This required fibers that could be processed more easily, such as wool or linen, and these again were linked to a settled life. In addition, the use of warp-weighted looms could only be developed by peoples who had developed beyond nomadism. The technique of weaving was developed during the Neolithic period, that is, the New Stone Age (in Central Europe, about 5500 BCE to 2200 BCE). Hand spindles and mobile weaving devices indeed also made it possible for even nomadic tribes to process fibers, and it is not to be excluded that nomadic peoples also decorated and dyed leather clothing, belts, and a few complex fabrics, but the professional dyeing of wool and linen, just like their processing, is associated with a settled life. This life made it possible to keep animals to produce wool as well as the complex processing of this wool, and thus associates the first comprehensive dyeing techniques as well as the spinning and weaving of complex fabrics with the Neolithic period.

Sophisticated weaving and dyeing techniques originated in Central Europe only in the Bronze Age, from about 2200 BCE, and impressive finds are documented especially from the Hallstatt culture of the eighth to sixth centuries BCE. Between 2002 and 2013, samples of sixty textile fabrics from the Bronze Age were analyzed, and we learned which dyes were used by the people from that period, and above all, what the dyes were extracted from. These samples from the Hallstatt period prove that, for example, it was already a tradition to dye textiles with dyer's woad. One of the most exciting discoveries of this time was how they extracted the dye from the plants and transferred it to textiles![1] It has not been proven whether the Celts dyed using fresh woad petals or used the so-called vat dyeing method (see p. 121 for an explanation)—in any case, the work consisted of extracting the indican dye from the plant and getting it into wool and linen. Indigotine is one of the insoluble pigments, which means that the dye can only be extracted by a chemical process, for example by fermenting with urine. This fact led to professional production of dye in so-called woad mills outside the towns during the Middle Ages—the smell of the fermenting woad must be rated as extremely unpleasant.

In addition to dyeing blue using woad, it could be demonstrated that the most important natural red textile dyes were already known in the Bronze Age. The red dye was extracted from so-called red plants, and indeed from the creeping roots growing underground. During the Celtic period, this meant using cleaver or bedstraw. It can't be clearly demonstrated whether the Celts already used dyer's madder. In parts of Europe, dyer's madder was already known as a dye plant at this time, but in other parts it only became known in the Middle Ages.[2] It's possible that red dyeing was done using a dye insect, the coccus cacti, but this can't be clearly established. For dyeing yellow, the Celts of the Hallstatt culture—that is, the period from about 800 to 470 BCE—already used reseda, also called dyer's weld, which is still common today. The dye recipes in this book are therefore based on an almost 3,000-year-old culture!

Whether the Celts used mordants containing metals to increase the durability of the dyes is not known; purely theoretically, however, copper and iron were available. Iron could also have been used to produce green hues. To do this, water that had been

1. Cf. Hofmann de Gejizer, p. 144.

2. Cf. Hofmann de Gejizer, p. 148.

Dyer's madder has always been an important plant for red dyes in Europe.

heated in an iron cauldron was poured over reseda-dyed fabric. Green dyeing using complex overdyeing techniques has, amazingly, been known since the Iron Age!

THE ROMANS: STRUCTURED, STANDARDIZED, AND STABLE

The difficulty of demonstrating how people did their dyeing during the Bronze Age and the Celtic period is mainly due to the fact that nothing was recorded in writing. While there are writings by ancient Greek and Roman authors about dyeing textiles, the first dye recipes only come from the Romans. The ancient Greek as well as the ancient Roman culture was, in comparison with that of the Celts, strongly oriented to philosophy, state law, and socially relevant ideas, which were also recorded in writing and thus available to posterity. No wonder that the Romans were convinced that they had developed the "true state."

There are legends of both gods and heroes relating to the discovery of several dyes, such as, for example, the story of purple. Purple is one of the most important and valuable dyes of antiquity, and according to legend, it was the Phoenician god Melqart who discovered the effect produced by the purple snail while walking by the sea. Supposedly his dog bit one of these snails in two, whereupon the animal's jaw turned purple. The god wiped the dog's mouth, believing that it was hurt, and found to his surprise that the cloth he had used was dyed with the most glorious red tone he had ever seen. According to this legend, dyeing with purple snails thus comes from the Phoenicians, and it has actually also been proven that the Phoenicians were using purple dye by 1400 BCE. This dye remained very popular among the Romans. There is a text from 301 CE that contains, among other information, the exact prices for purple. To convert the prices into today's currency is difficult, but it can be assumed from the literature that at least $9,200 in today's dollars was spent for slightly more than two pounds of purple dyed silk.[3]

But back to dyeing with plants.

Among the Celts, dyeworks still chiefly existed to supply their own needs, and therefore dying was done at home. Among the Romans dyeing was done commercially. There were separate workshops for each dye, the dyers were organized into guilds, and purple dyeworks were even under state authority. The first recipes with detailed instructions on dyeing come from the third century CE and involve dyeing with woad. Woad was regarded as the most important source of blue dye in the Middle Ages, before it was replaced by imported indigo. With the establishment of guilds and trade in precious dyed fabrics, the time of exclusively home-dyeing was also over in Central Europe. Dyeing rapidly changed from a self-sufficient craft to a trade that was subject to rules, and partly associated with a high level of turnover, and thus became more and more industrialized. The roots for this were already found in the regions outside Europe, among the Phoenicians, and in Europe during the golden age of the Roman Empire.

WOMEN'S OR MEN'S HANDICRAFT?

It can't be clearly proven whether and when dyeworks changed from a purely women's handicraft to one of men. But it can be assumed that dyeing materials remained in women's hands as long as it was still a home craft; thus, as long as it was for their own use, and at most, dyeing was done for small-time trade. There is no clear evidence on this topic. Occasionally there are attempts to determine, on the basis of the analysis of grave offerings, which

3. Cf. Ploss, p. 12.

activities should be assigned to which gender in pre-Christian times or in the first millennium CE. The remains of a burial mound in Norway are, for example, assigned to Queen Asa, who presumably lived between 800 and 850 CE. Many handcrafted objects such as weaving boards, hand spindles, and remains from dye plants were found in this grave, which has allowed some writers to be convinced that among the Viking of this time, dyeing was in women's hands and was even worthy of a queen. However, this can't be proven certainly. What is certain, however, is the fact that the dye plants found are woad and dyer's madder—the plants we still use for dyeing today.

It can be demonstrated that, at latest, with the organization into guilds by the Romans, the craft of dyeing was performed by men. By placing of certain guilds under state authority, the pressure was put upon the dyeworks to gain as much profit as possible—and this not first during the Industrial Age, but already in ancient Rome, where slaves worked in the dyeworks in factory-like mills!

THE MIDDLE AGES: CRAFT GUILDS AND STRICT CONTROLS

In northern European regions, the Middle Ages brought about the transformation of dyeworks from a home craft into a thoroughly organized commercial handicraft. Strict restrictions, varying from country to country and principality to principality, regulated dyeworks and trade in dyes. Dyeing was still done using plants, but in a much wider scope than during the time of self-sufficiency, small trade, and barter. With the rise of the cities in the Middle Ages, approximately between about 600 and 1500 CE, this brought about the departure from small communities that could provide for themselves. It brought much wealth to some people, but bitter poverty to many others. No professional area was excepted from guilds and trade guilds, which strictly regulated who could learn a trade—above all who could, after a long apprenticeship of up to seven years, the journeyman years, and production of a masterpiece, actually be designated as a craftsman and thus attain a little independence.

Dyers' guilds first existed in the sixteenth century, while south of the Alps, for example, in Florence, dyers were organized already in the thirteenth century into their own dyers' guilds. In Italy, they had access to dyes and dyeing techniques from the Orient much earlier on, while for a long time in Central Europe woad and dyer's madder were the main dye plants. In the Middle Ages, woad was the most important supplier of blue dye, but by using overdyeing, it was also possible to create other colors and color nuances using woad. The manufacture and, above all, the sale of woad were clearly regulated, and no one ever got by the guilds. The precious dye plant was exported from Germany to England, and raids on transports of woad were not rare. The example of woad also shows clearly how there were attempts already during the Middle Ages to offer dyes in compact form for trade: Dried woad was prepared in small balls for further sale. The demand for blue dye was so great that at the height of woad cultivation in Europe, entire villages and village communities could live just from its cultivation and processing. This high-time of woad ended with a dye that came from far-off India.

MODERN TIMES: FROM THE "DEVIL'S DYE" TO THE DEVELOPMENT OF SYNTHETIC DYES

At the end of the fifteenth century, the sea route to India was opened up, and as a result, large quantities of one of the most controversial dyes of history were brought from there to Europe: indigo. The dyestuff indica, intrinsic to the indigo plant, is the same as the dyestuff in dyer's woad, but in the indigo plant it comes in a significantly higher concentration. Indigo had been already known in Europe for several centuries, but transport of the precious dye by land was so lengthy, cumbersome, and expensive that indigo could not gain a foothold in traditional dyeing. With the opening up of the sea route, the scenario suddenly looked different, and the entirety of the woad dyeworks, a well-regulated medieval branch of industry, was in danger. Restrictions were the result. In the mid-sixteenth century, the prohibition of importing indigo was enacted in England, and the use of indigo was severely punished in many European cities. While in Frankfurt an official prohibition of indigo was enacted, the dyers in Nuremberg had to swear an oath every year not to work with indigo, and the penalty for dyeing with indigo was death.[4] Indigo, the "devil's dye," was a serious threat to the income source of many people in Europe and attempts to raise the plant in Europe failed. Indigo indeed thrives in Europe, but the plant here does not achieve the color results as we know them from Indian indigo.

All the restrictions of the late Middle Ages and the beginning of modern times could no longer stop the decay of the woad dyeworks; by the mid-seventeenth century, only a fraction of this once-blossoming economic sector remained, and the villages that were exclusively devoted to the production of woad had become fewer and fewer. By the mid-eighteenth century, indigo finally prevailed over woad. The higher quality of the imported dye could no longer be denied and the import prohibitions were lifted. At the time, no one could suspect the fact that the controversial dye from far-off India, after all the battles, would dominate the market for blue dyeworks for just a hundred years.

In 1826, the pharmacist Paul Unverdorbed distilled a colorless liquid from a piece of indigo, which he called "crystalline." At the same time, this discovery was made by a number of other researchers, and the material was thus given various names, including "aniline," and it was discovered that it was not only found in indigo, but could also be obtained in different ways. August Wilhelm Hofmann also found the substance in hard coal tar and thus prepared the way for the production of synthetic dyes. At the beginning of the nineteenth century, the battle between the synthetic blue and natural indigo was over, and after extensive research, it was possible to obtain a synthetic blue that was in no way inferior to the plant dye and could come out on top in terms of price.

In the history of mankind, dyes have always led to controversies, and dyes have always been of great importance. If it was a question in the case of purple, of who was allowed to wear the precious color, in the case of the color blue, it is very clear how industrialization began to determine the market by means of price. Each color has its meaning and its story in the history of mankind, and the respective way of manufacture contains secrets, and thus we can no longer bring many secrets to light today.

However, just as things already were during the long history of dyeworks, it is still so today: **Dyeing with plants** is an **art**, a **secret**, and an activity that **requires a holistic approach to the dyestuffs and the material**. No set dyeing results can be promised—the colors produced by the dyeing process depend upon the parts of the plants used, the harvest time, the condition of the material, and the various mordanting processes that can be done on the material. Thus, each plant dyer has their own methods and swears by them. The following descriptions of mordants and plants come from thirty years of experience, but they are certainly not complete—another person would dye differently and also achieve different colors. Anyone who is engaged in plant dyeworks will therefore develop their own ideas and their own recipes over time. Thus, the recipes offered here are also intended to be a reference point inviting you to experiment and develop further.

4. Cf. Ploss, p. 1.

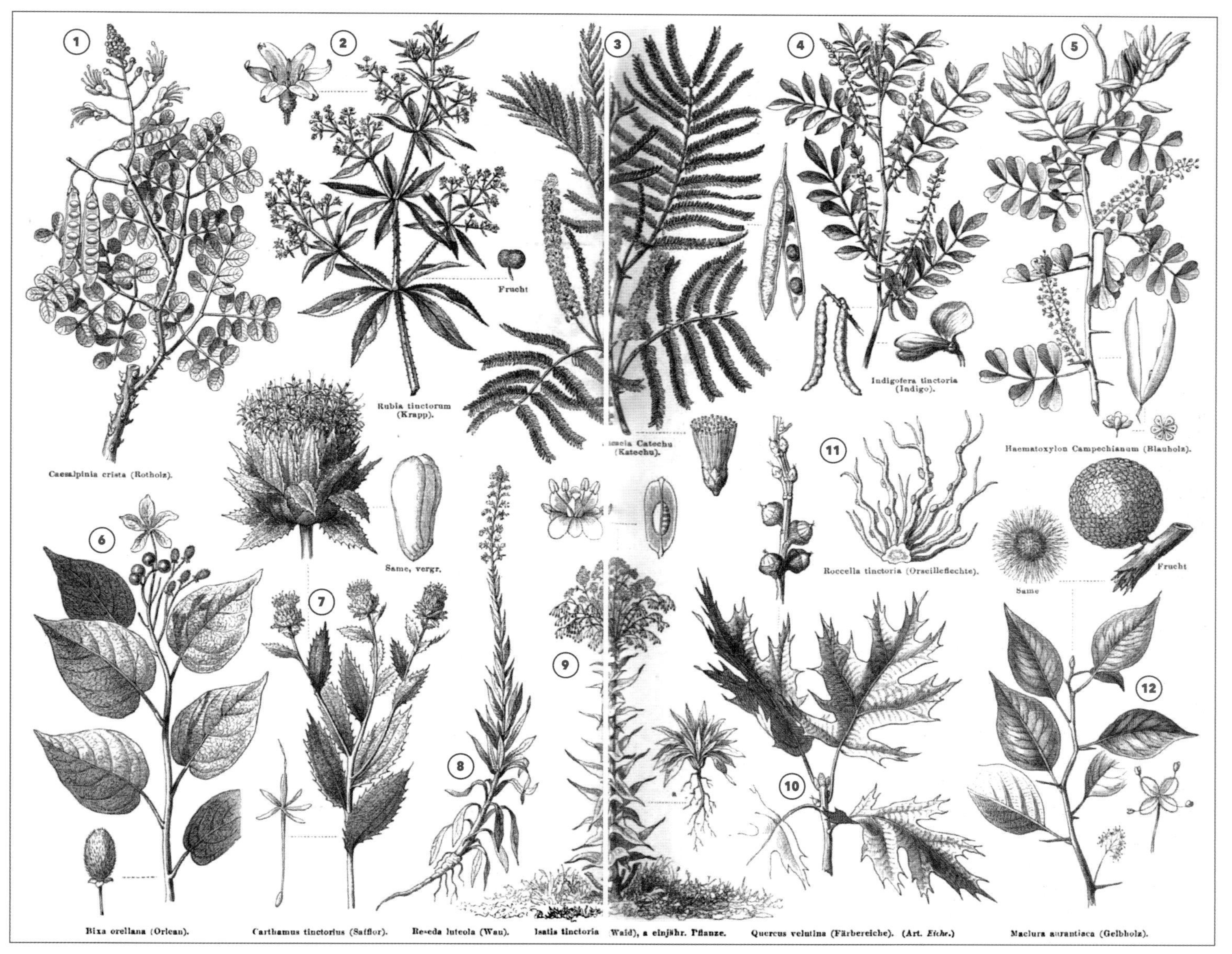

An overview of the most important dye plants of the nineteenth century: 1. Red wood or Brazilianwood (*Caesalpinia crista*); 2. Dyer's madder with flower and fruit (*Rubia tinctorum*); 3. Catechu, cutch tree, with flowers (*Acacia catechu*); 4. Indigo with flower and pods (*Indigofera tinctoria*); 5. Bloodwood with flowers and pod (*Haematoxylum campechianum*); 6. Achiote or annato with flower bud (*Bixa orellana*); 7. Dyer's thistle or safflower with flower (*Carthamus tinctorius*); 8. Reseda, also called dyer's weld (*Reseda luteola*); 9. Dyer's woad with flowers, seed, and young plant (*Isatis tinctoria*); 10. Black oak with acorns (*Quercus velutina*); 11. Orchella lichen (*Roccella tinctoria*); 12. Yellowwood or dyer's mulberry tree with seeds and fruit (*Madura tinctoria*). From *Meyers Conversation Dictionary*, 1888.

THE DYEING STUDIO

A Place of Color Magic

Everyone would like to have a dyeing studio—at ground level, with a panorama window and a view of the dyer's garden, as well as an open fireplace, with the copper kettle containing the dyestuffs hanging above it. Old crafts create archaic, romantic images in your head. If you set yourself on your own way with dyeworks using plants, you will quickly discover that while archaic images differ from present conditions, the magic of these things is not lost.

Setting up a dyeing studio is easier than you think. If you want to dye outdoors, a few containers and ingredients that you can store in a crate are sufficient. If you are dyeing in the house, you should consider whether to set up your own room, as dyeing often generates odors that you don't want to have in your own kitchen. In addition, of course, as with all craft activities with dyes and liquids, things can get sprayed and spilled. Dyeing processes require time, and in addition to hours of soaking or simmering, which require large containers, after-treatments and drying also take a long time. All this requires space and therefore you won't be happy in a small studio unless you plan things very well. In addition to this, you must use special pots, containers, and cooking spoons for the mordant and additives. The dyeing equipment must therefore be deliberately kept separate from your usual kitchen utensils.

There should in any case be a place where the containers of dye and textiles can be left to stand, as well as places where the dyed textiles can dry. For my dyeing studio, I have set up a large room in the basement where materials as well as dyestuffs can be stored and where I have space for the widest variety of dyeing processes. Anyone who dyes as a hobby can by all means fall back on the simplest equipment and dye outdoors—the results will be just as pleasing as those from a professional dyeing studio, and your own kitchen isn't permanently taken over. The odors produced during dyeing are thus banished from your own kitchen.

You can carry out all the processes described in this book using the following **basic equipment**:

- A stove or a burner
- For dyeing outdoors or indigo dyeing, it is recommended that you use an electric canning cooker with a thermometer, because the temperature should be maintained exactly.
- Pots of different sizes. In my collection of dyeing pots, for example, there are stainless steel pots in various sizes, so that even large amounts of material can be dyed comfortably. Pots from the family stock or from flea markets can also be used with confidence. For dyeing with bloodwood, a pot with rust spots is recommended, because the chemical process resulting from the rust reinforces dyeing the material black.
- Wooden cooking spoons in various sizes as well as other wooden tools like laundry tongs or Chinese chopsticks. Using these tools, you can take the textiles out of the pots during washing, mordanting, and dyeing without having to reach into the liquids with your fingers. I avoid metal tools, since these can always affect the dyeing process.
- Plastic buckets and vats in different sizes, at least one of these containers should be closeable with a lid. The cold mordant will be stored in this closeable container; the other buckets and vats are used for rinsing or temporary storage of dyeing fabrics.
- Sieves or strainers of different sizes, best made of enamel or plastic.
- Apron and rubber gloves
- A kitchen scale for weighing the dyestuffs. I recommend a postage scale for weighing the ingredients for dyeing with indigo.
- Dye bags for holding the steeped dyestuffs. For this, bags you sew yourself from an old plastic curtain, for example, work well. The material for these bags should not be too thick!
- A mortar, a hand mill, and a thermometer

- Clothes dryer and laundry pegs
- Cords, for example twine, and scissors
- Writing tools—you always get the best ideas and insights while you are working!

THE MATERIALS

Basically, all textile materials can be dyed, from wool to silk, cotton, or paper, to materials such as nettle fiber and hemp. From the history of the respective countries and regions, it turns out that the oldest dye recipes always refer to the fibers that were native to the respective country and usually still are.

Thus, in Central Europe there is a long tradition of dyeing with wool, but not with cotton, since this first arrived in Central Europe during antiquity. In

Basic equipment for the dyeing studio.

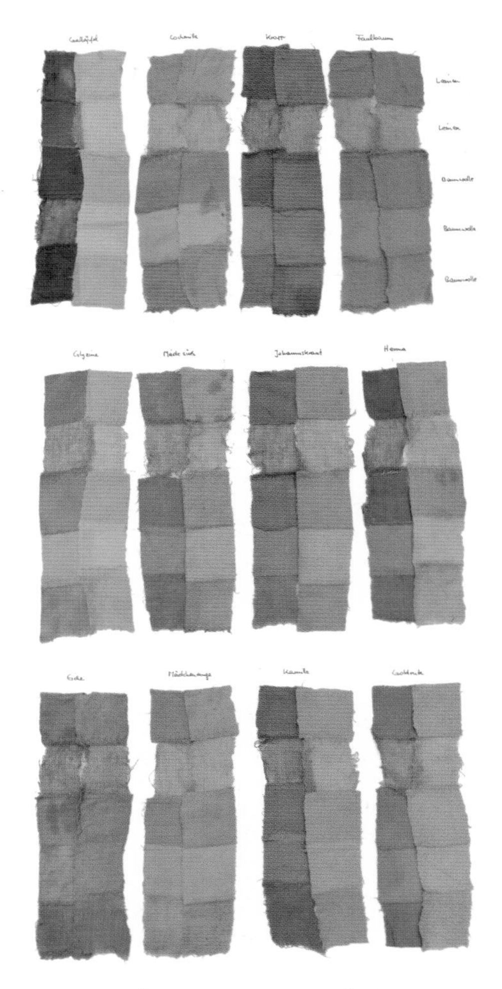

Linen samples—experiments in dyeing on linen and cotton.

the sixth century, cotton was already being grown in Spain, but it was still regarded as an absolute luxury good. The processing and thus also the dyeing of wool, on the other hand, has a very long tradition, as does the processing of linen. This material is well-known to us all from fairy tales, which often tell of "spinning flax." From antiquity to the European Middle Ages, linen was, next to wool and hemp, the most widely used material for clothing, but in the nineteenth century it was supplanted by the cotton fiber. Today, people are again increasingly recalling the quality of this native material.

Hemp fibers also have a very long tradition. The processing of hemp fibers into textiles was already known in ancient China, and findings go back to the third pre-Christian millennium. In Europe, hemp fibers can be found in graves from the fifth century BCE. The sturdy hemp fibers were used in the same way as flax and wool, and just like these, were supplanted in the nineteenth century by cotton. Today, hemp is again an up-and-coming fabric, and the properties of the hemp fiber are being compared with those of silk.

When it comes to dyeing textiles, we can see a fundamental difference in whether the materials are of plant or animal origin, and thus whether they consist of protein or plant fibers. In my long-time dyeing experience, my focus has been on protein fibers—that is, wool and silk—but I have also worked extensively with plant fibers. In my experience, protein fibers can be dyed better with most plant dyes than plant fibers can, but there are also exceptions.

From top to bottom: Silk, wool, nettle fiber, paper, and various raw materials in the dyeing studio.

WASHING

All the materials we want to dye should be washed beforehand. Well-washed fibers absorb both the mordant and the dyes better, resulting in more beautiful dyeing results.

Unspun wool in a fleece, carded top (pre-combed wool) or spun wool in a skein, as it has meanwhile become commercially available, is already pre-washed and free of all dirt particles as well as excess fats. If you buy these, you save yourself the long washing and thus also the detergent. However, if you want to buy untreated wool, you should wash it very well, since the material still contains a lot of natural fats. These prevent the color from being optimally absorbed. Neutral soap or a very mild detergent should be used for washing since this will protect the fibers. In principle, it is also possible to dye ready-spun balls of wool from the woolens shop. Here, you should make sure that you are working with a pure wool or wool with a minimal synthetic fiber content. In various dyers' blogs, there are examples of dyeing balls of wool with a proportion of synthetic fibers, some of which are listed in the appendix. You should also wash balls of wool from the store before dyeing. I myself only dye unspun wool and silk, because I continue to process them.

Unspun wool, unwashed and washed.

Washing Wool

For washing, fill a tub or simply the washbasin with as much water as it takes so you can move the wool around easily. It is very important that the wool doesn't get a temperature shock, so you should pay close attention that the water is warm to the hand. You should be very careful when adding the detergent—with pre-washed wool, a splash of wool detergent is enough. For completely untreated wool (unscoured wool), add a normal amount of detergent, as you can find from the information on the detergent package.

Immerse the wool gently in the water to avoid matting. You should not move the wool round too much; that is, don't submerge it too vigorously in the water since pushing it down heavily can cause matting. It is necessary to treat the material very carefully. Now, when the fibers are well wetted with the wash water, let the wool soak for about twenty minutes and then take it out and rinse it several times in clean water. My way of working is to carefully empty the water out of the container and refill it with fresh water. This

Fill a container with plenty of lukewarm water and add some detergent.

Carefully immerse the wool, press it into the wash water, and don't move it too much, to avoid felting!

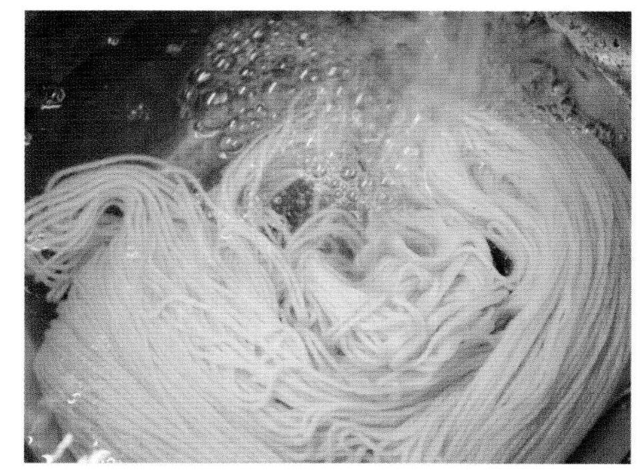

Rinse with clear water until no more detergent foams up.

process is repeated until you can't see any more detergent. The rinse water should be at the same temperature as the wash water.

Washing Silk

Silk is washed in the same way as wool, but silk is not as sensitive to the temperature. Nevertheless, silk should never be placed in boiling water, either during washing or dyeing, since the boiling process destroys the natural silky sheen. Fill a tub or washbasin with water, add a moderate amount of detergent, as for wool, and then you can immerse the silk cloths or the silk yarn into the wash water and move it around so that all the material is well immersed in the wash water. Some wild silk has to be washed very well, that is, with the amount of detergent indicated by the manufacturer on the package; since silk bast (silk gum, a glue-like protein that surrounds the raw fiber) and dirt residues are found in wild silk. When rinsing silk, you do not have to be as careful with the fibers as with wool, since silk does not get matted. However, in any case, you should take care to ensure that the final rinsing water is completely clear and contains no detergent residues.

Washing Plant Fibers

In contrast to animal fibers, when washing plant fibers it is less a matter of extracting the dirt that has accumulated on some of the fibers, and more a matter of getting out the fibers' own dyestuff found in the fiber. Thus, for example, when washing out nettle fibers, the wash water becomes very reddish to brown, and a dirty yellowish when washing out cotton. But please understand that this is only true of fibers left raw, which you can hardly obtain commercially today. The production of cotton is industrialized to such an extent that we can no longer get any natural cotton fibers commercially. This also applies to hemp fibers and linen. Should you have the luck to find really untreated, natural plant fibers, then these should definitely be washed well to get out the fiber's own color. I have found raw nettle fiber at a dyestuff market and would like to describe the process based on this fiber.

Fill a pot with plenty of water so that the fibers are well covered, and add a tablespoon of soda and a tablespoon of detergent for five liters (1.32 gallons) of water. Then stir well once, immerse the plant fibers, bring the wash water to boiling, and then simmer it over a low flame for several hours. Then let the water, with the fibers in it, cool down and let it soak for another few hours. If the wash water is strongly colored, the process has to be repeated. This may have to be done up to five times. Washing plant fibers can, depending on how natural they are, take a long time! The difference with already pre-treated materials is only that we use considerably fewer chemicals and certainly less aggressive detergents in washing and dyeing in our home studio.

With silk, there is no risk of felting, so it can be vigorously immersed in the wash water.

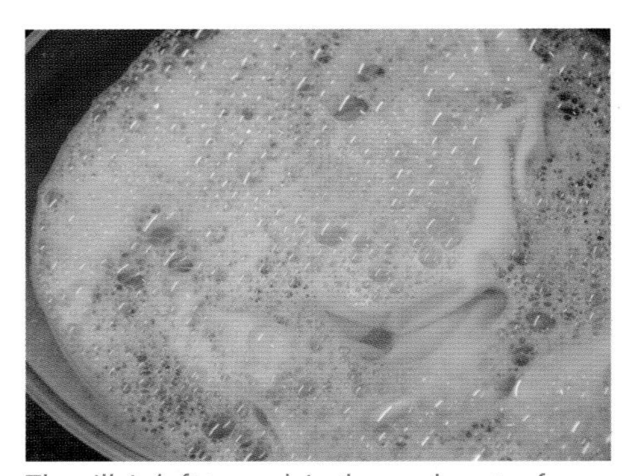

The silk is left to soak in the wash water for about twenty minutes; wild silk, for about thirty minutes.

Carefully rinse out the detergent residue with fresh water.

Cotton fabrics can definitely be pre-washed in the washing machine, just like canvas or hemp. As already mentioned, however, these fabrics are usually thoroughly pre-treated when they are purchased commercially.

NOTE

Essentially, we can still say that almost all ready-woven fabrics, whether of plant or animal origin, can be dyed with plant dyes. Similarly, these fabrics must be well washed, best in a washing machine, since most of the commercially available textiles are pre-treated and processed. Furthermore, it should be noted that all textiles dyed with plant dyes can't be machine washed, but are only hand washable.

THE MORDANT

Mordanting is a chemical process necessary to prepare the textile fibers for the dyeing process. Mordanting can be done using various agents and in different processes, namely by applying heat during the mordanting process or by so-called cold mordanting.

The most common mordanting agents include alum, iron and copper sulfate, tin chloride, and potassium dichromate.[5]

5. Cf. Prinz, p. 3.

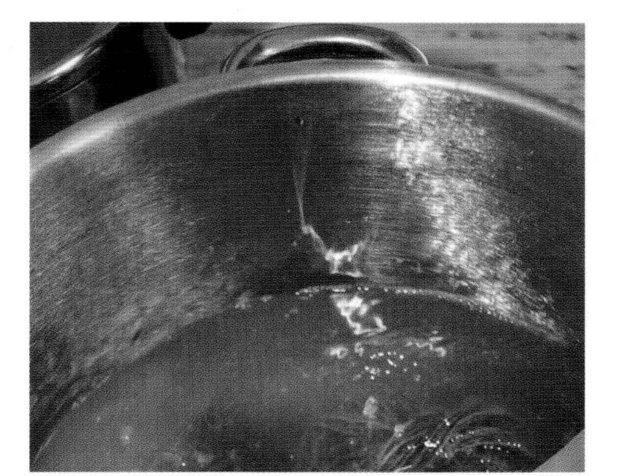

Add the plant fibers to the pot with water, detergent, and soda.

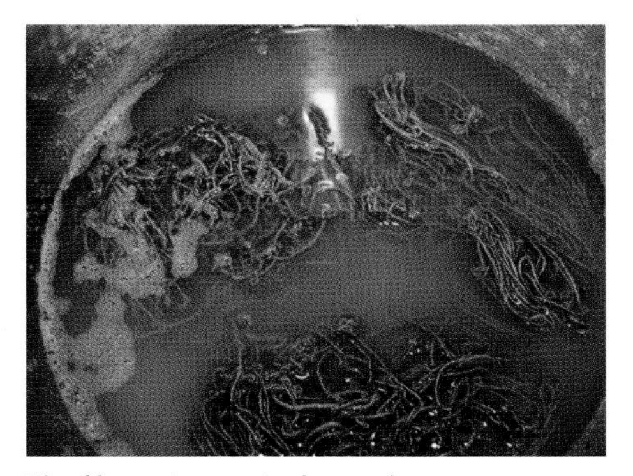

The fibers simmer in the wash water.

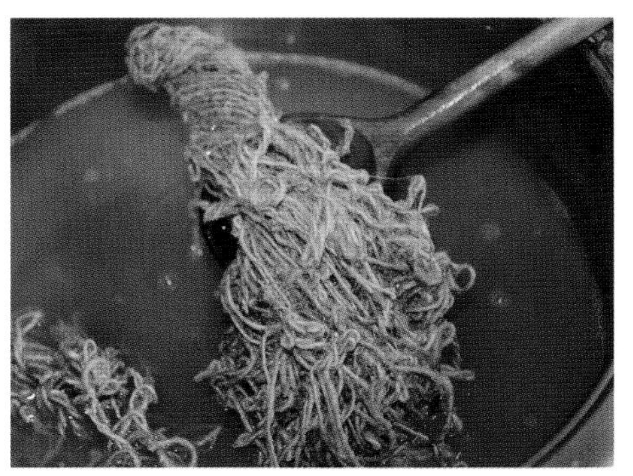

As long as the wash water remains very discolored, the process has to be repeated.

These mordants have an influence on the color development in the fabrics, but it is not only the different mordants that are responsible for different shades. The materials from which the pots used for mordanting are made also affect the dyes. This knowledge is as old as the use of metal kettles. While copper pots or iron pots that are used in mordanting have a substantial effect on color development, pots made of enamel or stainless steel are color neutral. Thus, already during mordanting, there are countless ways to affect and to design the dyeing process.

NOTE

The secret of mordanting is simple to explain: Mordanting activates a chemical process, which makes the fibers more receptive and thus allows the colors to become more intense. However, like any chemical process, the fiber will be partly damaged by the mordant, so it is important to be very careful, especially with respect to temperatures. This will be explained in more detail in the section "Pre-mordanting" (next page).

The ingredients for mordanting: From alum to iron water to potash.

The essentials for mordanting: Large pots, gloves, and wooden tools.

In addition to mordanting agents that are dissolved in water, you can also use plants that release chemical substances for the mordanting process. This approach has been handed down from the Middle Ages and is no longer common today, just like mordanting with urine. The plants that can be used for mordanting include club moss, bearberry, or chickweed, which are rich in potassium. However, these plants were always used in combination with suitable pots, since it was discovered already at the beginning of the history of mordanting how different plants react with different metals. It was also known that tannin-containing plants that are boiled in an iron kettle replace the iron sulfate. Plants that contain tannin include, for example, sorrels, oak galls (also called oak apples), and the black alder.

Nowadays, pre-mordanting is done with alum, an aluminum mordant. From a scientific point of view, alum is a "water-containing sulfate potassium-aluminum compound," which was already known as a mordanting agent by the Romans. Alum is found in nature in volcanic rocks, for example. Today alum powder is available in pharmacies and from online suppliers. Adding tartar to the alum mordant protects the fiber of the textiles and makes the colors more intense. Iron sulfate is usually used for post-mordanting, but copper sulfate is also used in post-mordanting.

Pre-mordanting

Pre-mordanting is done before dyeing and is intended to make the textiles dye-absorbent. Pre-mordanting is done using either alum or aluminum cold mordant; both are available in specialty stores as well as in pharmacies. For mordanting with alum, the mordanting agent is first dissolved in warm water. The best method is to add the mordant agent to a glass jar with a screw top, add warm water, close the jar, and shake it until the agent has dissolved. Pour the dissolved mordant agent into a pot and add as much water as you need so you can move the dyeing fabric around freely.

Then immerse the already dampened textiles in the cold mordant and slowly—over one hour—heat it to 194°F (90°C). This careful heating process protects the fibers and prevents the wool from matting, for example. After reaching 194°F (90°C), leave the fabrics to soak in the mordant for one hour at this temperature and turn them over twice during this time, so that the mordant also reaches all fibers evenly. Afterwards, you can by all means leave the fibers soaking overnight in the cooled mordant.

Even if you do this mordanting processes very cautiously, the textile fibers are still damaged, since even 194°F (90°C) will stress all textiles. While I have used alum as a mordant since I began my dyeing activities, for some years I have been using aluminum cold mordanting, since this protects the fibers and also simplifies the dyeing process. It is thus possible to mordant sensitive fibers such as silk and wool gently, since the fabric does not have to be heated.

However, if you want to intensify red dyestuffs by adding tartar, alum remains the only option because it is the only mordant that works with tartar. Therefore, for cochineal, dyer's madder, and for some yellow dyes such as turmeric (higher proportion of red in the yellow color), I still use the classic mordant of alum and tartar. This is indicated separately in the respective dye recipes.

IMPORTANT

When handling mordants, always wear rubber gloves and make sure that there is no food in the vicinity of the mordanting process. In addition, all mordant agents and dyeing accessories should be kept out of the reach of children. Even if we use the most natural agents possible and mordant as gently as possible, these substances can cause skin irritation and should not be taken internally!

Alum Mordant

For mordanting with alum, in general, mordant the dyed material by weight by using a 10% portion of alum by weight. For example, for 1 kg of wool, prepare 100 g of alum. As already explained, dissolve the alum powder in a glass jar of water, empty this into a pot of hand-warm water, immerse the dyeing fabric in the liquid so that it is completely covered, and slowly bring it to a boil. The dyeing fabric should simmer slowly for at least an hour, then cool slowly, after which it can be removed from the mordant. For cooling, the fibers can also be left in the mordant overnight.

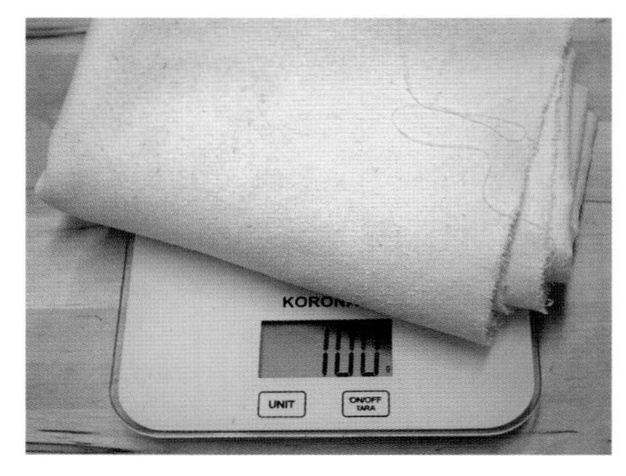

Weighing the dyeing fabric.

10 g of alum is required per 100 g of dyeing fabric.

Mix the alum powder with water in a glass jar, close it, and shake it very well until the powder has dissolved.

Pour the contents of the jar into a pot with water and stir thoroughly, then add the dye and boil.

IMPORTANT

To intensify the colors, you can add tartar to the alum mordant, which yields wonderful results with reddish tones. Here, as with the alum, the amount of tartar used should be 10% by weight of the weight of the dyeing fabric; that is, again, 100 g of tartar for 1 kg of wool. This is added to the alum mordant before this mixture is added to the mordanting water.

Cold Aluminum Mordant

For cold aluminum mordanting, dissolve 100 g of mordant agent in warm water, pour this into a bucket, and fill it with up to five liters (1.32 gallons) cold water. Stir this liquid well, then you can immerse as much of the dyeing fabric as will still be covered by the liquid.

The dyeing fabric must soak in the cold mordant for at least three hours, after which it should be thoroughly rinsed and then dried, or immersed in the plant dye while still damp.

The material can also remain soaking in the mordant for a longer time; this won't change the results any more. After taking it out of the mordant, rinse the material thoroughly and wring it out carefully before immersing it in the dye bath.

Simply cover the mordant with a lid. Cold aluminum mordant can be used again and again, in contrast to other mordants; in fact, until there is no longer enough liquid left to cover the dyeing fabric.

Dissolve 100 g of aluminum cold mordant in some warm water.

Pour the dissolved mordant into a bucket and dilute it with 5 liters (1.32 gal.) of water.

The dyeing fabric is immersed in the liquid for at least three hours and moved around a little so that it is completely covered.

Take the dyeing fabric out of the mordant and rinse it carefully so that no mordant remains in the fabric. Cover the mordant bath; it can be reused!

IMPORTANT

When preparing cold mordants, gloves and possibly mouth and nose protection should be worn. As long as the powder hasn't come into contact with water, it is slightly corrosive. As soon as the aluminum powder is dissolved in water, there is no longer any danger of a chemical burn.

Tannin Mordant

Tannin is a vegetable tanning agent that occurs, for example, in the oak gall (also called oak apple). Tannin is used for pre-mordanting plant fibers. To do this, use 20% by weight of the dyeing fabric, that is, 200 g of tannin per kilogram of textile. Dissolve the tannin powder in water, and use as much water as it takes to cover the dyeing fabric. Bring this mixture to a boil and keep it simmering for two hours. Then turn off the stove and leave the dyeing fabric in the liquid, preferably overnight, or at least twelve hours, and longer if possible.

After this process, as before, dissolve 20% of alum and 5% of soda in hot water; that is, calculated for 1 kg of dyeing fabric, 200 g of alum and 50 g of soda. Immerse the fabric, well-rinsed material from the tannin mordant, in this mixture, bring it to a boil again, then keep it at a boil for an hour.

Leave the dyeing fabric soaking in the cooling liquid, preferably for at least twelve hours. After this process, rinse out the dyeing fabric and then either dry it and store it or dye it right away.

Tara Powder

Mordanting with tara powder is another possible way to pre-mordant plant textiles. Tara powder is extracted from the *Caesalpinia spinosa*, a tannin-rich plant from the St. John's wort family. Tara is applied in the same way as tannin, but it has a stronger effect and leaves a preliminary color, so that it is only used for darker colors.

Dissolve the tannin powder in water.

The dyeing fabric is slowly brought to a simmer.

Cooling down in the tannin mordant.

Pre-mordanted textiles react strongly and immediately to tara in combination with iron and take on a much darker color. Thus, for example, to create a dark gray coloration, add iron water to the tara mordant. In this case, the mordant bath is also the dye bath.

Post-mordanting

Post-mordanting offers possibilities for color nuancing and for creating darker colors. For the post-mordant process, the dyed textiles can be dried before they are given the next process, but this is not necessary. I definitely find it practical to continue working with the wet dyeing fabric. Post-mordanting is done with using iron sulfate, iron water, or potash, all substances that are available in dyer's stores or pharmacies. I myself treat the dyed fabrics only with self-made iron water or potash; substances such as copper sulphate should not just be discarded in waste water and should therefore be avoided for home use.

For post-mordanting with **potash**, dissolve one tablespoon of the substance per 500 g of dyeing fabric in warm water, add this mixture to a bucket with sufficient water to completely cover the dyeing fabric, and immerse the dyed textiles in it. Let the dyeing fabric soak in this post-mordant until it has taken on the color that you would like to obtain for your result. This can potentially take an entire night.

It is also very easy to make **iron water** and it works wonderfully. For the iron water, add 100 g of small pieces of iron (nails, etc.) to a glass jar with 400 ml of water and 100 ml of 25%-strength vinegar essence. This has to stand for two to three weeks until rust forms and darkens the water. As soon as the solution has become opaque, it can be strained and used. You can wash and dry the pieces of iron and store them until the next time you make iron water. You can also keep any iron water you don't use mordanting so you always have iron water in stock and ready at hand when you need it.

Post-mordanting with iron water is very simple. Before the process, take the dyeing fabric out of the dye water. In the pot with the dye water, pour a good shot of iron water. Stir the mixture thoroughly and again immerse the dyeing fabric in the dye bath, the dye liquid. If necessary, you can stir carefully again or warm up the liquid again to enhance the darkening effect. As soon as the textiles have taken on the color tone that you imagined, take the textiles out of the post-mordant. Rinse the dyeing fabric well after the post-mordanting process!

Put small pieces of iron in a glass jar with vinegar and water.

The prepared iron water in a jar.

Pour the unused iron water into a bottle for storage and label it.

DYEING

Dyeing with plants is very different from dyeing with industrially produced dyes. Even if you don't plant, harvest, or dry all the plants you want to use yourself, these plant dyes bring a different energy into the dyed textiles than synthetic colors do. The application is more complex, usually takes longer, and the color results are not as consistent as with industrial dye. However, anyone who has deliberately compared a plant-dyed fabric with one that is industrially dyed, will be enchanted by the color intensity and brilliance of the plant dye.

Dyeing Methods

You can dye using flowers, roots, leaves, and fruits and by overdyeing, that is, a second and third dyeing process after applying the first dye. In this way, you can create essentially any hue from the color palette. Experimenting with different mordanting and dyeing methods and with post-mordanting yields the most varied results, and in this way, each plant dyer develops their own colors and color nuances as well as their own recipes. The various possibilities for dyeing depend in part on the plants you use, and in part they are matters of taste. These methods will be presented in the following pages, and each time we will note along with the individual plants which methods are to be used for dyeing.

A layer of nut shells.

Firmly press down a layer of spun wool.

A layer of wool, unspun.

Add a layer of nut shells. Repeat this as many times as you wish. At the end, add a layer of nut shells, and press this layer down firmly once again.

Cover the contents of the pot with cold water and leave to soak for a week.

After a week, take the wool out of the water and remove the shells. Small pieces will fall off after drying. The wool changes color somewhat during the drying process!

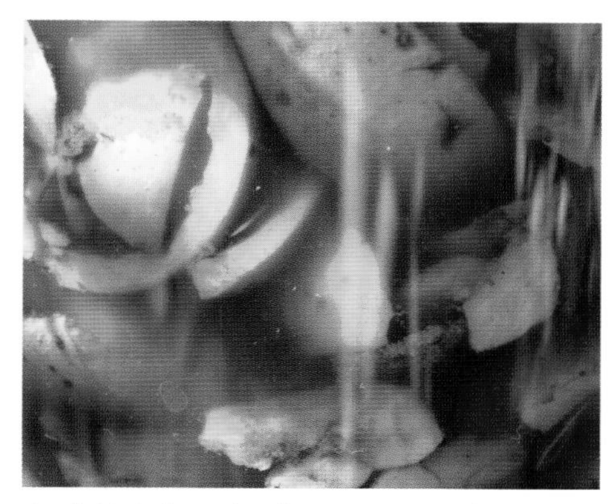
And this is how the dyeing process looks in a glass container after a week.

Cold Dyeing

Cold dyeing is the simplest dyeing method. Cold dyeing is suitable, for example, for dyeing with green nut shells. The dyeing fabric is layered with green nut shells in a container (bucket, wash tub, etc.). The more shells you use, the more intense the colors; basically, you can assume you need about 5 kg of shells for 1 kg of material. The textiles should be left in the container for at least one week, then take them out and hang them up to dry. Dyeing with nut shells does not require any pre-mordant, since the nut shells are rich in tannic acid.

Direct Dyeing

For direct dyeing, the dyestuffs, that is, woods, are soaked, preferably overnight. The next day, boil them for at least an hour and then strain. The dyeing fabric is then immersed in the liquid, slowly brought to a boil, and dyed by boiling slowly for about an hour.

The dyestuff, in this case pieces of oak wood, is soaked overnight.

The next day, the dyestuff is strained into a dye bag.

Tie the bag tightly and put it back into the dye bath.

Add the dyeing fabric to the dye bath and stir gently, so that all the fibers are well covered.

This form of dyeing works well for all plants that contain a lot of tannins, such as oak bark, walnut shells and leaves, lichens, or chestnut leaves. There is no need for a pre-mordant because of the high tannin content of these plants.

Single-bath Dyeing

This dyeing method also saves the process of pre-mordanting and works well with birch leaves, onion skins, and alder buckthorn bark. The dyestuffs are soaked overnight, boiled and strained, as for direct dyeing. Next add the mordant agent (alum) directly to the dye bath, and likewise the amount should be 10% by weight of the dyeing fabric. This means that 100 g of alum is required for 1 kg of textiles.

Then add the textiles to the dye bath, heat this slowly, and boil slowly for an hour. Then let the dyeing fabric cool in the dye bath, remove it, and rinse it out well.

Contact Dyeing

Soak the dyestuffs overnight and then boil for an hour, but don't strain them out afterwards. Add the fibers to the dye bath, where they come into direct contact with the respective dyestuff. There are also variations that require the dyeing fabric and the dyestuff to be heated together. Which variation works well for which plant is described in the individual recipes.

It certainly can be very tedious to extract the small pieces of plant from the wool, but the dyeing results often speak for themselves. It is a good idea to simply try both contact dyeing and direct dyeing, and decide for yourself which color results are better.

Dyeing Pre-mordanted Material

Through the pre-mordanting process, the fibers become more absorbent of the dyes, resulting in brighter colors. Pre-mordanting has already been described in detail (see page 20). After mordanting, the textiles must be rinsed well and can then be dyed.

For this dyeing process, soak, boil, and then strain out the dyestuff; only then are the textiles added to the dye bath. Precise information on these dyeing processes, that is, the length of soaking time for the dyestuff and how long the dyeing fabric should remain in the dye, can be found in the corresponding recipes.

Dyeing in Steps

Dyeing in steps can be used both for direct dyeing and contact dyeing.

Dyeing in steps improves the lightfast quality of the dyed textiles. Take the dyeing fabric out of the dye bath after half an hour and hang it up directly in the open air without rinsing. After another half an hour, return the dyeing fabric to the dye bath. This operation can be repeated easily. Here, it is again true that you should decide for yourself which color result is the one you desired.

NOTE

With all these possibilities it quickly becomes apparent that it makes a lot of sense to document your first attempts at dyeing. A small booklet with comments on your best dyeing results can help you achieve exactly the color you want the next time, without having to experiment all over again. At the same time, as when working with any natural materials, it will never come out just the same—nature always helps a bit in deciding things!

The Basic Dye Recipe

For beginning dyers, a simple basic recipe, which can be used for all dyes, should serve as a basis. This basic recipe is a point of reference—we note changes to the recipe for the individual dye plants that achieve particular results as well as special cases that affect how you handle the plant.

Basic Recipe

For the basic recipe, use 1 kg of dyestuff per kg of dyeing fabric; for example 1 kg of goldenrod for 1 kg of wool.

Soak the dyestuffs overnight in a large pot, in as much water as you need so that the dyestuff can soak up well and the dyeing fabric is well covered. The next day, heat the dyestuff slowly, bring it to a boil, and boil it for one to two hours.

Then let the dyestuff cool in the water, and finally strain it out.

Immerse the (pre-mordanted) dyeing fabric in the dye bath and heat again slowly. Depending on the material, boil it for about an hour and then let it cool slowly. Then remove the dyeing fabric and rinse it well, so that it can be either dried or immediately post-mordanted. The dye bath can be used again; it is possible to use it for further dyeing. With each dyeing, the dye bath becomes somewhat weaker and the results become lighter accordingly.

Rules of Thumb

You can advance from the basic recipe to a better level of understanding of dyeing technique, if you consider the structure and quality of a plant. You get the most colorfast, most lightfast color results when you dye using **roots** and **root barks**. The plant's color is strongly bonded in the root and you must first use various processes to dissolve it out. We don't see the color beforehand; the color is only revealed while the dyestuff is boiled and when the dyeing fabric is soaked in the dye bath. The root's quality is the quality of earth.

Above ground, the plant has the quality of water; the color is now already partly visible and easier to extract from the plant. The results of **dyeing using the whole plant** remain very colorfast and lightfast, but in comparison to root dyes, you should already expect some slight fading. This increases when you change to flowers.

Color results when using **flower dyestuffs** are often surprising; for example, when you use dahlias, you can obtain very intense colors, but these colors are much more transient than if you dye using the whole plant. The air element, which is represented by the flower, is strong here—the color results are not always visible and the colors fade faster. This does not mean that textiles dyed with flowers lose their color after just a few days, but you cannot avoid changes in color due to sunlight.

A plant's color reaches its peak in the **fruit**. Here we see the color in its full splendor; think of blueberries, elderberries, or pokeberries. If we get fresh fruit on our clothes, they stain quickly and visibly, without further exposure to water or chemistry. The fruits of a plant represent fire energy; here, the plant has arrived at the end of a process, and the cycle of life begins again from the start, with the seed contained in the fruit. Dyeing results are intense when you use fruits, but in comparison the least colorfast.

The parts of the plant also differ enormously in terms of the amounts of dyestuffs. If you use the same amount by weight of roots compared to dyeing fabric, this is still a manageable quantity. When using dried plant parts, you need significantly more in terms of quantity, even if it is still the same in terms of weight, thus the same amount of dyestuff as dyeing fabric. For most flower recipes you need at least twice as much flowers as textiles. To obtain good color results with fruit, it takes four to ten times as much fruit by weight. That is why I don't use fruits for everyday dyeing, because it is better to process quantities like this in jam jars than on textiles. For this book, however, a few recipes with fruits were chosen to show that it does work.

But back to the plant and its qualities. Just as the elementary qualities of earth, water, air, and fire exist in the plant, these qualities are also necessary for a dyeing process: The earth, that is, the minerals, is added by means of mordanting; the water serves to dissolve the dye together with the heat, that is, the fire; and the color results become visible only through drying in the air.

The Rules of Thumb for Dyeing

- You get the most colorfast dyes from roots, then from the whole plant, followed by the flowers, and finally the fruit.

- Flowers often produce very different results. Here it is worthwhile to carefully note the time of harvesting, the location and appearance of the flowers, if you want to get similar results again.

- To be able to dye using plants, you always need all four elements—thus, for cold dyeing or solar dyeing, you also need the sun. Also, to dye without using a mordant, you at least need the minerals themselves in the dyestuff.

- Drying in the air brings out the final color result—if you still aren't sure about a color, then it is worth it to let a piece dry before taking all the pieces out of the dye bath.

THE DYER'S GARDEN
Where Color Magic Blooms!

A dyer's garden with some classic native dye plants is a feast for the eyes and also a paradise for bees and other insects. These native dye plants include woad, dyer's madder, reseda, dyer's broom, yarrow, and many more. Most of the plants native to Central Europe will dye in various shades of yellow and brown; woad is the famous blue dye plant, and madder dyes red.

It doesn't matter if you have space enough to plant the entire herbal color palette or a small selection in your home garden, dye plants enhance your own garden. It is also worth including a dye plant in your repertoire that you don't use to produce any dyestuff. For example, extracting the blue dye from woad takes an effort that is hardly worth it, but the plant thrives in your own garden. The same is true of the intensely flowering indigo: It enhances every garden, but you would need entire fields full of indigo to extract the dye and in addition compensate for the fact that the indigo plants growing in our latitudes do not provide the same quality as the plants growing in India.

We can't be certain whether people differentiated clearly between a kitchen garden and a dyer's garden in the Middle Ages or earlier, but the fact is that so-called wash and dye plants were not allowed in any home garden. Although dyeing was strictly regulated in the Middle Ages, we can assume from this that dyeing was still established as home craft, at least in the countryside, and only gradually disappeared from everyday life due to industrialization.

Tickseed, blooming beauty from the dyer's garden.

SUITABLE PLANTS FOR THE DYER'S GARDEN

The dye plants that I cultivate in my own dye garden include dyer's broom, wild indigo, indigo, dyer's chamomile, alkanet, dyer's madder, and safflower. Reseda (dyer's weld) requires a dry location and therefore does not grow everywhere. While it doesn't want to thrive in my garden, it grows very well at my girlfriend Gerti's under the balcony, and I can look forward to a rich harvest every year. Anyone who dyes using plants knows that it is never possible to grow everything in one garden, and therefore you have to work with other dyers, neighbors, and friends by networking and exchanges. This book should also be a call for you to intensify such connections. While researching some recipes that I did not have in my repertoire, for example, I made contacts that made my further work much more fruitful.

As far as a dyer's garden is concerned, you can also let yourself be influenced by other dyers, gardeners, garden lovers, and their gardens.

You can cultivate a colorful variety of dye plants in your garden and extract brilliant dyes from them.

The following garden flowers, for example, are good for dyeing and have their place in the dyer's garden:

- Hollyhock
- Dahlia
- Mexican marigold
- Common marigold
- Rose
- Tickseed

The following plants from the range of medicinal herbs play a role as dye plants:

- Goldenrod
- St. John's wort
- Yarrow
- Horsetail
- Lady's mantle
- Stinging nettle
- Tansy

The following plants from the vegetable garden can be used for dyeing:

- Onions
- Rhubarb
- Carrots

The leaves of some trees are also suitable for dyestuffs:

- Birch
- Walnut
- Elder
- Tanner's sumac
- Apple tree

The indigo plant in my own garden.

In the next chapter I will describe my dye plants. I have divided them into plants from the garden, those from the forest, trees, and dyestuffs from the kitchen and from afar.

DYER'S GARDENS TO VISIT

After you start to get interested in dyeing with plants, you'll inevitably want to get to know the plants more closely. Many living history museums and botanical gardens offer well-arranged herb gardens and also dyer's gardens. Usually beautifully designed, they can serve as an incentive to plant your own dyer's garden. When you travel, it's worthwhile to make side trips to visit other regions' dyer's gardens.

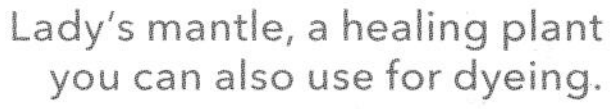

Lady's mantle, a healing plant you can also use for dyeing.

Larkspur.

Hollyhocks.

DYE PLANTS

From the Dyer's Garden

In the following pages we will present the most common dye plants that grow in our dyer's gardens. These include plants native to the Central European region, and plants that have become indigenous with us or at least grow under our climatic conditions, even if their relatives in distant lands may be better suited for dyeing, as is the case for indigo. You can obtain most of these plants already prepared as dyestuffs in specialty stores—dried and crushed wood, herbaceous plants, or leaves in crushed or ground form. It is up to each dyer to decide whether to harvest and dry the dyestuff themselves or rather to use the purchased product. In this book we note about most plants whether it is worthwhile to harvest the roots, leaves, or flowers yourself, because this is clearly not just a question of the time involved but also a question of the quantity you have available. If, for example, plenty of dahlias grow in your neighborhood and you are allowed to pick then, then it makes sense to use them for dyeing. But if you have to tediously wait for the next single bud to bloom, it is better to choose a different plant.

There are sayings and legends about many plants that we should not neglect, and the same applies for information about the healing or health-promoting effect of each plant. We will also go into the locations for each plant, what makes it easier to gather and harvest some of them, as well as a small guide for cultivating them yourself.

For all those plants that I use in my daily dyeing work, I describe additional dyeing recipes as well as provide samples of the color on various materials. The reason that I don't use all the plants described for dyeing, is that I have found out for myself, in my long years of experience, which ones work best and also the most colorfast dye recipes, which I have integrated into my daily work. After a few years, every dyer will have selected out precisely those plants and dyes that give the best results on the chosen material. Because the material is also a determining factor in the results!

The marigold, a vibrant joy for any garden.

Dyeing samples, carefully labeled.

While I mainly dye silk and wool myself, other dyers have specialized in using linen or cotton and have developed their own recipes. Once again, this is a reason to document your dyeing results well during the first years of your life as a dyer, by keeping a dyeing diary and photographs. This way, you can be sure to preserve the best of your own recipes.

Many plants from a native dyer's garden color in yellow hues of all shadings, so it is scarcely possible to use all these plants in your daily work. While many dyers swear by marigold blossoms, for example, I myself do not use this plant for dyeing. You can also obtain good yellow results with the sunflower, from which you can use the petals, and likewise with oregano or sage. You can also dye yellow tones with mullein.

ALKANET, DYER'S OX-TONGUE

(ALKANNA TINCTORIA)

DYE SAMPLES

Wool

without mordant

alum/tartar

cold mordant

cold mordant/ iron water

Silk

without mordant

alum/tartar

cold mordant

cold mordant/ iron water

The *Alkanna tinctoria* comes originally from the Mediterranean region, but can also be grown here. For all those who are interested in science, please note: All plants that have the additional designation *tinctoria* in their Latin name work very well for dyeing! Naturally, plants without the designation *tinctoria* can also be used for dyeing. Like all Mediterranean plants, alkanet loves warm and dry conditions, and it loves locations where it is not directly in the rain. Like reseda, *Alkanna tinctoria* grows in such places as cracks at the edge of the path and on house walls, where it stubbornly defends its place. The Romans already used the dyestuff from the alkanet root for makeup, together with beeswax, since dyestuffs in *Alkanna tinctoria* are fat soluble.

For dyeing textiles, we use the *Alkanna tinctoria* roots. You can gather and dry these yourself, but this is somewhat laborious, since the dye is only found in the root bark. Therefore, you need a lot of roots. For 100 g of wool, you have to use at least double the amount of dried alkanet root, and it should be noted that the dyestuff in the bark is not very water soluble. If you want to get the purple hue hidden in the plant, you first have to pour medicinal alcohol over the dyestuff and then use the alcohol extract in the dye bath. However, I don't use this pre-treatment, because I create purple by overdyeing. Basically, dyeing with alkanet works quite simply if you dedicate some time to it. With the dyeing recipe I use, you obtain the widest variety of gray shades to a grayish aubergine.

Alkanna tinctoria

DYE RECIPE

For my dyeings, I soak the alkanet root bark for two or three days in water—twice as much dyestuff by weight as dyeing fabric. By this soaking, I get the dyestuff to slowly wash out.

After the soaking, I boil the roots for at least two hours. Then I strain out the root pieces, put them in a dye bag, and back into the dye bath. I heat the pre-mordanted fabrics very carefully in this dye bath for about an hour to obtain beautiful brown-gray tones.

For post-mordanting, take the dyeing fabric and dyestuff out of the lukewarm dye bath and add a shot of iron water to this. The amount of the "shot" varies by the weight of the dyeing fabric; for 100 g of wool, half a shot glass is enough, for 1 kg of wool, about half a wine glass. Here it is worth noting that using any more iron water does not achieve anything—it doesn't enhance the effect! Put the dyeing fabric back in the dye bath mixed with iron water until you get the desired color result; tones from dark gray to gray-green and aubergine gray are possible. If the result is not dark enough, you can again gently warm the dye bath.

In my experience, alkanet dyes wool to clear shades of gray, while it always results in a light brown gray on silk.

Alkanet root for dyeing.

Wool and silk, dyed with alkanet.

WILD INDIGO

(BAPTISIA TINCTORIA)

DYE SAMPLES

Wool

1st dyeing
cold mordant

2nd dyeing
cold mordant

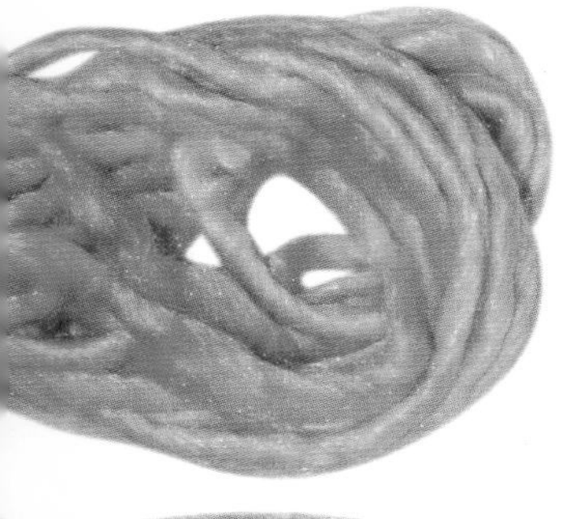

Silk

1st dyeing
cold mordant

2nd dyeing
cold mordant

The baptisia, also known as blue wild indigo or false indigo, originally comes from North America, but it has become an ornamental plant in Central Europe. The plant prefers sandy, dry soils and catches your eye with its wonderful blue-violet flowers. As a dye plant, the name "false indigo" is somewhat misleading, because the plant that grows here, the *Baptisia tinctoria*, dyes pre-mordanted wool green. You cannot dye any blue tones with the indigenous wild indigo!
Native North Americans knew that they could dye clothing blue using *Baptisia australis*, but no recipes remain for these dyes. As also from the Celtic period in Central Europe, there are no written records from native North Americans, and much of their old knowledge will probably remain buried forever.

Once again, for baptisia: Here any first-time dyer can keep experimenting happily! The color results from our self-harvested plants depend very much on their location, the weather, and the time of harvest, so that sometimes you can expect surprising results.

DYE RECIPE

To dye with baptisia, I pick fresh leaves from the branch—double the quantity in proportion to the dyeing fabric—then tear them in pieces and boil them for at least an hour. Then I let the dye bath cool with the plant pieces in it. Also

Blue wild indigo.

for the dyeing process itself, leave the plant pieces in the dye bath; here, the dyeing occurs so to speak through contact. Immerse the dyeing fabric, pre-mordanted with a cold mordant, in the dye bath and dye it while boiling for one hour. For wool, please be sure to carefully increase the temperature slowly! After one hour, let the dye bath cool with the dyeing fabric in it, then dry the dyeing fabric and extract the remains of the plant. Then wash out the textiles—the results are light- to medium-green tones!

I proceed in the same way for the second dyeing, but I leave the dyeing fabric in the dye bath until it has taken on the desired color tone. This can range from dark yellow to green on the color palette, each depending on how much time you give the dyeing fabric. Simply try it once, leaving the dyeing fabric overnight! Since the dyeing fabric takes on the green color over time anyway, I omit post-mordanting with iron water for baptisia.

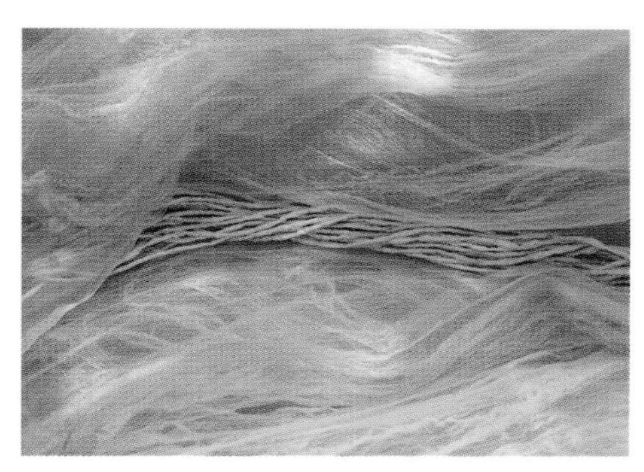

Silk dyed with baptisia.

Wool and silk dyed with baptisia.

DAHLIA

(*DAHLIA* SPP.)

DYE SAMPLES

Wool

1st dyeing
cold mordant

2nd dyeing
cold mordant

1st dyeing
cold mordant/
iron water

Silk

1st dyeing
cold mordant

2nd dyeing
cold mordant

1st dyeing
cold mordant/
iron water

Dahlias bloom in the most beautiful colors in almost every garden. Originally these aster family flowers came from South America; they are native to the highlands of Mexico and Guatemala, and arrived in Europe in the eighteenth century. Dahlias became so popular in Europe during the past centuries, that now festivals and exhibitions are held and the colorful blooms not only decorate gardens but balconies as well. The plant blooms from summer to autumn, and you use the flowers for dyeing. It is recommended to harvest these before the edges of the flowers have already turned brown in autumn, so that the petals still have their dyeing power.

The dahlia works wonderfully well for dyeing, but the color results are extremely different. The flowers just don't dye the way they look. It is therefore necessary to enjoy experimenting if you want to dye using dahlias, since it is hardly possible to determine beforehand which color results you will get. For example, I have obtained a bright orange from some yellow-red flowers, which I hadn't expected at all!

DYE RECIPE

Boil fresh dahlia flowers, at least double the quantity of the dyeing fabric, for an hour. Then let the dye bath cool down and strain out the

Fresh dahlias in a dehydrator.

A small miracle in orange: wool and silk dyed with dahlia.

flowers. Put the cold-mordanted dyeing fabric in the dye bath for about an hour and boil carefully. You can expect yellow and orange tones as a result. After dyeing, take the dyeing fabric out of the dye bath. For post-mordanting, add a shot of iron water to the dye bath and again immerse the dyeing fabric in it. You can get yellow-green and dark green shadings of color this way—take the materials out of the dye bath when the desired hue is obtained. As for all dyeing processes: Drying in the air will again change the color a bit. Each dahlia species also reveals its own individual shading during dyeing; here, this is worth researching and making your own experiments.

It is also possible to dye using dried flowers and this again creates different shadings of color. Drying the flowers is very simple: just lay the flowers loosely on a tray; if you have large quantities, a sheet also works well. Then let the flowers dry well before storing them in paper bags or cartons. As for all dried dyestuffs, it is very important that the plants are completely dry, since otherwise they can become moldy and all your effort was in vain. Furthermore, the dried flowers should be stored either in large jars of dark glass, in a cupboard or in a paper bag, but anyway so that they are not constantly exposed to light. If in the autumn you are still getting large quantities of flowers, you can also dry these in a dehydrator.

Dyeings of silk and wool done with dried dahlias.

DYER'S THISTLE, SAFFLOWER

(CARTHAMUS TINCTORIUS)

DYE SAMPLES

Silk

without mordant

alum/tartar

cold mordant

Wool

cold mordant

alum/tartar

The safflower originally came from Asia, but was already known by the 3rd millennium BC to the Egyptians, and was brought to Central Europe by the Romans. The plant has been indigenous to Central Europe since the thirteenth century. From time immemorial, the safflower has been used not only as a dye plant, but the plant seeds are also used to make oil.

The safflower is also called "false saffron" because its flowers were used as a substitute for the much more expensive saffron. There is a long history of obtaining dyestuffs from the flowers of the aster family. With the safflower, it is quite easy to obtain a rich yellow by cold dyeing fabric that has not been pre-mordanted. Saffron dyeing works consistently on plant materials much better than on materials that contain proteins. In the specialist literature it is noted that safflower will also dye fabrics rose color and even pink, but there are different instructions for this in circulation—I have written down those which I found most useful. The basis for the potential to use the safflower to dye fabrics in red hues, lies the fact that the flowers contain red dyestuff as well as yellow. But it is not as easy to extract the red dyestuff from the dyer's thistle as it is for the yellow, and it requires some practice to extract this dyestuff.

Safflower.

Safflowers in full bloom.

DYE RECIPE

In addition to cold dyeing with safflower, which can extract both colors from the plant and which is described in the next recipe, it is also possible to hot dye with saffron according to the basic recipe. The coloring results range from yellow to yellow-green to olive green, by post-mordanting. The red dye can't be dissolved out from the plant through this conventional dyeing process; it remains concealed from us.

For dyeing according to the basic recipe, use the same amount of dried safflower blossoms as of dyeing fabric, and soak these for one hour, after which they are boiled for an hour. Then strain out the flowers and slowly reheat the dyeing fabric in the dye bath and dye it for an hour. By post-mordanting with iron water, you can create olive-colored green tones. To do this, take the fabric out of the dye bath, add a shot of iron water to it, and then re-immerse the dyeing fabric in the dye bath. What is striking for me when dyeing with safflower is that the yellow tones are more intense on wool than on silk, but the green tones also become more intense on silk by post-mordanting.

Cold dyeing with safflower

For me, the easiest method to extract both dyestuffs from the safflower is the following: Soak 100 g of safflower in 3 liters (0.79 gallons) cold water and let it stand for a few hours; this also works if you let the dye bath stand overnight. During this time, the yellow dyestuff dissolves out of the flowers. After being allowed to steep, filter the flowers into a dye bag; the yellow dye bath will be used for yellow dyeing and is set aside!

Wool and silk dyed with dyer's thistle.

Silk

without mordant

alum/tartar

cold mordant

Linen

without mordant

without mordant

Now wash the flowers in the dye bag under running water until no more yellow dye is released. Then put them into a new pot with fresh, cold water, about 2 liters (0.53 gallons). Now add a teaspoon of baking soda and knead the flowers thoroughly—this process extracts the red dye from the flower. Take your time for this process and go on kneading the flowers for up to half a hour or longer. After this kneading process, the dye bath will already be dyed red, so now strain out the flowers.

To lower the pH value of the dye bath, add a shot of vinegar, immerse the dyeing fabric in the dye bath, and leave it to soak there for two or three hours. Finally, add vinegar once again, to lower the pH value to about 6. The best way to test this is by using a test strip from the pharmacy. Now you can leave the dyeing fabric in the dye bath, as desired—take it out from time to time, wring it out, and check whether the color corresponds to what you had imagined. After you reach the desired color, you now still have to rinse out the dyeing fabric. Then dissolve a tablespoon of citric acid in 3 liters (0.79 gallons) of water and add the dyeing fabric to this bath for about one hour, to fix and intensify the dye. If you leave the dyeing fabric in this solution for too long, the pink becomes quite light blue. The results from my own attempts have resulted in pale pink to pink (on linen)—so here prospective dyers will be challenged in their enjoyment of experimentation!

Wash out the safflower blossoms under running water.

In practice it's been shown that plant materials take on the red tones from the safflower blossoms much better than animal fibers do; the best is to use linen fabrics, which revealed themselves after the dyeing process in a wonderful fresh pink. The results on silk ranged from pale pink to skin color; personally, I found the tones on wool unusable. The great master Yoshioka shows, however, that dyeing with safflowers also creates wonderful results on silk. In his remarkable book on dyeing, *The Rainbow Color Thief*, he presents his personal dyeing results with safflower—of course with his very own recipe.

Linen fabrics dyed pink with safflower.

Cold dyeing with safflower, pink.

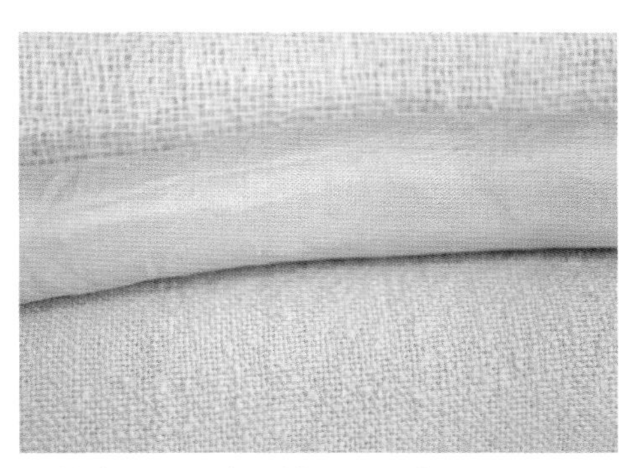
Cold dyeing with safflower, yellow.

Cotton and linen dyed with safflower.

DYER'S GREENWEED

(*GENISTA TINCTORIA*)

DYE SAMPLES

Wool

1st dyeing
cold mordant

2nd dyeing
cold mordant

1st dyeing
cold mordant/
iron water

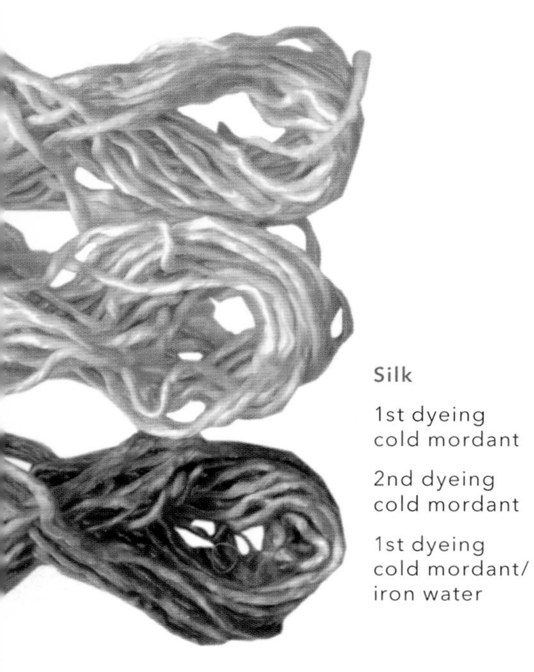

Silk

1st dyeing
cold mordant

2nd dyeing
cold mordant

1st dyeing
cold mordant/
iron water

The dyer's greenweed has been known as a dyeing agent for a long time, and is also known in medicine, but the plant is classified as "mildly poisonous," and it is not advisable to experiment with teas or extracts. In folk medicine, the plant was known for flushing out the system, and dyer's greenweed is also used in homeopathy.

In the Middle Ages, dyer's greenweed was the most important plant for dyeing yellow, especially in combination with overdyeing using dyer's woad. The green that is extracted went down in history as "Kendal green," named after the northern English textile town. Even Shakespeare is said to have described this color in his play *Henry IV*.

But back to the gorgeous yellow we get from the dyer's greenweed! The whole plant can be used for dyeing. Cut off the twigs and leaves and crush them; these can then be immediately used for dyeing. The plant can also be dried—the color results are the same.

For drying, lay the entire plant out on a cloth or hang it up in the air. Do not

Dyer's greenweed.

Wool dyed with dyer's greenweed.

dry dye plants in the blazing sun! As soon as the plant is thoroughly dry, cut it up and store it in a paper bag or carton.

DYE RECIPE

If you want to dye using the fresh plant, use double the amount of dyestuff in proportion to dyeing fabric; for dried plants, use the same amount. Thus, for 100 g of dyeing fabric, use 200 g of fresh dyer's greenweed or 100 g of dried. Soak the dried dyestuff overnight; if you are using fresh plants, one to two hours is enough. Then simmer the dye bath for one to two hours. Strain out the dyestuff, and after that proceed according to the basic recipe. The pre-mordanted dyeing fabric should be boiled on a low heat for one to two hours in the dye bath. The color results range from sun yellow to luminous light yellow tones from the second dyeing. By post-mordanting with iron water, you can obtain dark green and olive green. To do this, take the dyeing fabric out of the dye bath and add a shot of iron water to the still-lukewarm dye bath. Then put the dyeing fabric back into the dye bath and take it out when it has reached the desired hue. What is striking for me about dyeing with dyer's greenweed: While the yellow tones come more intensely into their own on wool, the results of post-mordanting come out better on silk. Another area for experimenting!

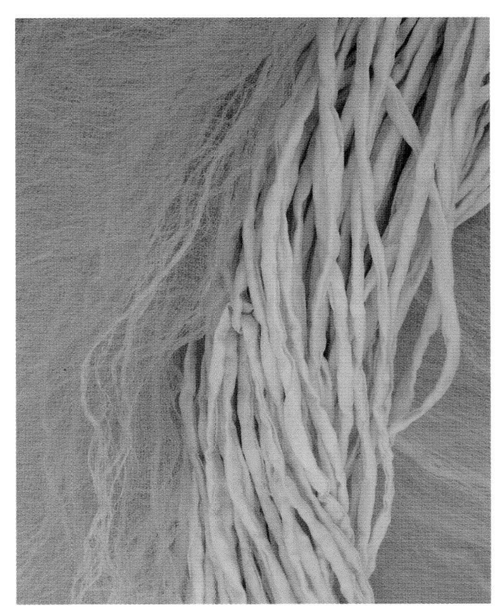

Wool and silk dyed with dyer's greenweed.

DYER'S CHAMOMILE

(*ANTHEMIS TINCTORIA*)

DYE SAMPLES

Wool

cold mordant

cold mordant iron water

Silk

cold mordant

cold mordant iron water

The dyer's chamomile grows in native dyer's gardens and produces gloriously sunny yellow tones. Originally the plant comes from the Mediterranean area, but it also thrives beyond the Alps without any problem. The yellow flowering plant is related to chamomile but has no healing effect. However, the dried flowers have a wonderful effect in herbal tea mixtures, since the dyer's chamomile not only dyes textiles, but also colors the tea yellow.

Dyer's chamomile flowers can be used fresh for dyeing, but I have found in practice that you seldom have that many flowers available at once. It is a good idea to harvest and to dry a few flowers every day throughout the summer—then in autumn you have gathered enough dyestuff to bring a summer yellow to your dyeing textiles. For drying, the flowers are simply laid on a cloth, but don't leave it in the blazing sun. The flowers dry quickly and can then be stored in a paper bag, a glass jar, or a box until you want to use them for dyeing.

DYE RECIPE

For dyer's chamomile, we may proceed

Dyer's chamomile.

Dried dyer's chamomile.

roughly according to the basic formula; thus we use the same amount of dried dyestuff as dyeing fabric. Soak the chamomile flowers for a few hours and then boil them for an hour. After that, let the dye bath cool down and strain the flowers into a dye bag. Put the dye bag back in the dye bath, bring this slowly to a boil with the dyeing fabric, and leave the dyeing fabric in the dye bath until the desired hue is reached. As with most recipes, it is also possible to do second dyeings here, which means that you use the dye a second time and thus obtain lighter colors. Post-mordanting with iron water also works very well and delivers a rich olive green. For this, after taking the dyeing fabric out of the dye bath, add a shot of iron water to the lukewarm liquid and then put back the pieces of dyeing fabric that you want to dye green. It is a good idea to stay by the dye bath and watch the pieces, because the dyeing fabric must be taken out of the dye bath when the desired green tone is reached. This tone changes quite quickly during post-mordanting, and it can happen that the result becomes too dark if the dyeing fabric is not removed in time.

Dyer's chamomile is also particularly good for overdyeing with dyer's madder, with which I obtain very beautiful and intense orange tones. For overdyeing with dyer's madder, immerse the still-damp yellow-colored material in the madder dye bath and leave it there until it reaches the desired color. For overdyeing with dyer's madder, it is the case that any material left in the madder dye bath too long just becomes red, so that the madder dye bath completely overdyes the yellow hue, but on the other hand, proceeding the opposite way does not produce the same results. Thus you don't get any such beautiful orange tones if you first dye with madder and then dye yellow on top.

Silk dyed with dyer's chamomile.

LADY'S MANTLE

(ALCHEMILLA VULGARIS)

DYE SAMPLES

Wool

1st dyeing
cold mordant

2nd dyeing
cold mordant

1st dyeing
cold mordant/
iron water

Silk

1st dyeing
cold mordant

2nd dyeing
cold mordant

1st dyeing
cold mordant/
iron water

Lady's mantle, also called parsley piert or lady's bedstraw, is another of the most important medicinal plants in folk medicine. Lady's mantle has probably as many names as healing powers, and in every region this herb is given other nicknames. Lady's mantle is administered for acute inflammation, and is likewise used as a blood cleansing agent. It also has its place in the field of gynecology as a remedy for menstruation and menopause symptoms—and this has been the case since ancient times.

There are countless legends and stories about the dewdrops that the plant "sweats" through its leaves. The alchemists tried to make gold from the dewdrops, while the Germanic peoples were convinced that lady's mantle dew came from the tears of Freya, the goddess of love and fertility. Because of these drops that are excreted from the plant, the lady's mantle also served as a "weather plant." If there are drops on the plant, this means it will rain. In addition to its health aspects and mystical stories, the lady's mantle is a dye herb that gives us a bright yellow, and with post-mordanting with iron water, yields all shades of gray tones with a very light impression of green.

DYE RECIPE

Once lady's mantle has established itself well in your dyer's garden, you will always have it available in abundance. That is why I personally dye using only fresh lady's mantle. I pick the leaves before it flowers, when the herb garden is covered

Lady's mantle.

Wool dyed with lady's mantle.

with a small carpet of lady's mantle leaves. For dyeing, as with most fresh plants, use double the quantity of dyestuffs in proportion to the dyeing fabric. Soak the leaves for an hour in a large pot and then boil them for one hour; strain them and then the dyeing process can begin.

Heat the pre-mordanted dyeing fabric slowly and boil it for one to two hours in the dye bath, and the result is a delicate, bright yellow. The results of the second dyeing come out brighter and they remind you of a very rich sun yellow. To obtain gray tones, take the dyeing fabric out of the dye bath, treat this with a shot of iron water, and return the dyeing fabric to the dye bath. The dye bath is still lukewarm. Leave the material in the dye bath until the desired gray coloring has been reached. This can range from a greenish gray to dark gray tones; it depends strongly on how long you leave the dyeing fabric in the post-mordant. Dyers who like to experiment can venture with warming the dye bath once again, which makes the results even darker. Dyeing with plants should make us try new things and not fixate on already existing recipes!

Preparation for dyeing.

DYER'S MADDER

(*RUBIA TINCTORUM*)

DYE SAMPLES

Wool

cold mordant

alum/
tartar/

alum/
tartar/
iron water

alum/
tartar/
potash

Silk

cold mordant

alum/
tartar

alum/
tartar/
iron water

alum/
tartar/
potash

Dyer's madder is one of the oldest known dye plants and was already known in ancient Egypt as well as among the ancient Greeks and Romans. The root of this rather unimpressive plant dyes textiles to a dark orange-red, the so-called madder red. Since extracting the dyestuffs from roots takes a lot of effort, it is advisable in this case to make use of the commercial dyestuff. The dyestuff of the madder plant is found in the root bark, which must be washed, cut, and dried well. Only this drying process can release the red color. Dyeing with fresh madder roots will therefore not yield the desired madder red.

You can get root pieces commercially, chopped small and dried, and store them until needed.

Madder was traditionally used as a healing plant, especially for symptoms of urinary complaints and renal diseases. In the meantime, madder root remedies are no longer approved, because some ingredients are considered to be harmful to health. However, this only applies to taking it internally; as a dyeing agent madder is completely harmless.

DYE RECIPE

When I dye with madder, I use the same amount of dyestuff as dyeing fabric. Soak the dyestuff in water overnight and heat

Madder roots.

Spun silk dyed with madder.

it slowly the next day. The dye from the madder root dissolves particularly well in soft water, so using rainwater is optimal. If you don't have any rainwater available, a small shot of vinegar helps. For dyer's madder, it is important to make sure that the dyeing fabric is not heated to more than 176°F (80°C); otherwise you won't obtain the desired brick red when dyeing, but rather a brownish red. After the overnight soaking, heat the dyestuff very slowly and keep it simmering for two to three hours, but the water should not boil at all; it should be kept lower than 176°F (80°C). Let the dye bath cool until only warm to the hand.

Then filter the madder roots into a dye bag. Put this in the dye bath together with the pre-mordanted dye, since dye will still dissolve out of the root.

Heat the dyeing fabric slowly to almost 176°F (80°C) and leave it in the dye bath until the desired color is reached. With dyer's madder it is a good idea to let it take enough time, because the dye gets transferred to the dyeing fabric slowly. The dyeing process may take several hours overall. After the first dyeing, it is possible to do further dyeings and the color nuances can then range to a very bright apricot. We can't provide precise information here—what is needed is the dyer's enjoyment of experimenting! For the last dyeing, you can also definitely leave wool in the dye bath overnight, but this is not recommended for scarves, as they can get spotted. I obtain the best coloring results by pre-mordanting wool with alum and tartar as well as pre-mordanting silk in an aluminum cold mordant.

Various materials dyed with madder.

TICKSEED

(*COREOPSIS* SPP.)

DYE SAMPLES

Wool

cold mordant

alum/
tartar

cold mordant
iron water

Silk

cold mordant

alum/
tartar

cold mordant
iron water

The tickseed or coreopsis designates an entire genus of plants. These annuals are characterized by especially lovely, delicate petals, which range from light yellow to dark yellow to pink, depending on the species. Originally the plant comes from the North American continent, where it is still found today in the wild, while in Europe the plant has found its place as a garden flower. The tickseed is also available as *Coreopsis tinctoria*, dyer's coreopsis, but not only this species will dye; in principle all the coreopsis plants that we can obtain here will do so. In contrast to most other plants that have become indigenous to our gardens, the coreopsis has no tradition as a healing plant, except in its original home where the roots of the tender plant were used as tea for diarrhea.

Just the lovely look of the tickseed alone is enough to appeal to any hobby gardener's heart, but in the dyer's garden, it opens up completely new possibilities for the color palette: The tickseed yields very intense orange tones. The coreopsis is very undemanding in terms of soil and thrives in almost every garden. There are both annual and perennial varieties. For a dyer's garden, it makes sense to plant perennial varieties and to let a few flowers bloom until the seeds are ripe, to make sure you always have seeds available. Both plants and seeds are available in nurseries, and it is simple to grow yourself. The self-harvested seeds also work well without

Tickseed—beautiful in every garden.

Dyed wool alongside fresh flowers.

any problem. The only weak point of this delicate coreopsis is probably the fact that slugs love this plant. Thus, if you find a lot of these voracious slimy creatures in your dyer's garden, it's better to grow the tickseed in pots. There it will bloom all through the summer, and you can harvest the fresh flowers before they wither. Leave some seeds to harvest from the plant. The harvested flowers are dried, since you will never obtain a sufficiently large quantity to be able to dye using fresh flowers, unless you plant an entire field, which would probably be a classic feast for the eyes.

DYE RECIPE

For dyeing with tickseed, use the same quantity of dried dyestuff—and that means only the flowers—as of dyeing fabric and soak these overnight. In the morning, boil the dyestuff for an hour, then filter it into a dye bag, and re-immerse in the dye bath together with the pre-mordanted dyeing fabric. Heat the dye bath carefully and simmer the dyeing fabric in it for a half hour to an hour. You can take out the dyeing fabric as soon as it has reached the desired hue.

When dyeing with tickseed, it is particularly striking that you will obtain very different results on wool, depending on whether you have pre-mordanted with a cold mordant or with alum and tartar. With the alum and tartar pre-mordant, a bright orange shade will emerge, but when you use an aluminum cold mordant the results come out much more subdued. On silk there is not such a big difference between the two pre-mordants. As is the case with most dyeings, it is possible to do a second dyeing, which yields fine apricot tones. Post-mordanting with iron water also creates wonderful results. For this, add a shot of iron water to the dye bath after dyeing, and immerse the already-dyed dyeing fabric in the dye bath again. Depending on how long the post-mordanting lasts and also on the material, you will get various hues of brown. On wool the color is rust brown, and on silk the brown tone gets a light tinge of green. It is worthwhile to experiment, because the color results are very intense and appealing in all their nuances!

Wild silk dyed with tickseed.

RESEDA, DYER'S WELD

(RESEDA LUTEOLA)

DYE SAMPLES

Wool

1st dyeing
cold mordant

2nd dyeing
cold mordant

1st dyeing
cold mordant/
iron water

Silk

1st dyeing
cold mordant

2nd dyeing
cold mordant

1st dyeing
cold mordant/
iron water

Dyer's weld is a biennial plant and comes originally from Asia. Dyer's weld thrives wonderfully in the Mediterranean region, and the main cultivation areas in the Middle Ages were in southern France and Italy. But reseda also thrives on the other side of the high Alpine ridge and has been used as a dyeing agent here for a long time before the Common Era. The entire plant is used for dyeing. This should be harvested when the flowers have already formed seed pods but these aren't yet mature enough to drop off during harvesting. The small seeds of the dyer's weld contain most of the dyestuffs for its bright yellow dyes. The plants must be dried well after harvesting, but never in the sun since this diminishes the color quality. After drying, chop the plants up small and store them in a paper bag or carton. Here, too, avoid direct exposure to the sun. Furthermore, it is important to make sure that the dried plant pieces are not exposed to moisture; a dry cellar or a cupboard makes the best storage place.

Reseda, together with madder and woad, is one of the oldest dye plants in the European region and completed the color palette in the Middle Ages. Madder was used for red tones, woad for blue tones, and reseda for a bright yellow.

Reseda finds its way through every small crack—once it's well established, it stays put.

DYE RECIPE

Soak dried, chopped reseda herb (all the plant parts dye), preferably overnight, and after soaking, boil for at least an hour. Here, too, use the basic quantities, the same amount of dyestuff as dyeing fabric; that is, you need 1 kg of reseda for 1 kg of wool or silk. After boiling, let the dye bath cool, strain the reseda into a dye bag, and put it back in the dye bath. Now add your selected pre-mordanted dyeing fabric to the dye bath and boil it slowly again until the dyeing fabric has taken on the intense reseda yellow. This takes about a half hour to an hour. During this time, stir the dyeing fabric around carefully in the dye bath and regularly check whether the desired color has already been reached.

For me, the bright reseda yellow is the least influenced yellow; it doesn't show any tinge of green or brown, but rather a strong, bright sun yellow. The clear reseda yellow also works the best for overdyeing with indigo, so you can obtain all the nuances of the green palette. You will only get the clear reseda yellow with the first dyeing, since all subsequent dyeings become very light. For the subsequent dyeings, you can let the dyeing fabric cool in the dye bath after one hour of boiling, and even overnight.

Wool dyed with reseda.

Dried plant parts on dyed wool.

DYE SAMPLES

RHUBARB

(*RHEUM RHABARBARUM*)

Wool

1st dyeing
cold mordant

2nd dyeing
cold mordant

1st dyeing
cold mordant
iron water

Silk

1st dyeing
cold mordant

2nd dyeing
cold mordant

1st dyeing
cold mordant
iron water

Anyone who has grown up in the country might still recall: Grandma used to peel the fresh rhubarb stems, cut them into pieces, and then they were dunked in sugar and eaten. Sour! And sweet at the same time! Rhubarb is a popular cultivated plant in Europe, but comes originally from the areas near the Himalayas. In the eighteenth century, the plant came to Europe via Russia and was first cultivated in France, the Netherlands, and Great Britain. If the plant is grown in a well-watered, medium-heavy soil, it is relatively undemanding, and sun to semi-shade are enough for good growth. Rhubarb is mainly made into jam and compotes or added to pie and cake. Rhubarb incidentally is a vegetable, although, because of the most popular ways it is prepared—as compote or jam—many people assign it to the fruits.

While the thick, fibrous leaf stalks are consumed, the rhubarb root is used as a remedy and also as a dyestuff. Dried, powdered rhubarb roots are used as a remedy for heartburn, constipation, and diarrhea, as well as for stomach and intestinal catarrh and liver and spleen symptoms. The root should be harvested from plants that are at least five years old.

For the dyestuff, I dig out my excess rhubarb plants. In old gardens you often find an impressive stand of

Freshly dug rhubarb roots.

Wool and silk dyed with rhubarb.

rhubarb, where it doesn't matter if you take some of the root stocks. Rhubarb propagates very well, and if you don't thin it out from time to time, it takes over the garden after a few years!

DYE RECIPE

For dying with rhubarb root, you need twice as much of the dried roots as dyeing fabric, that is, 2 kg of root pieces for 1 kg of textiles. A dehydrator works particularly well for drying the roots. Soak the crushed, dried root pieces for two to three days. Then boil the root pieces until they are soft.

In my experience, this takes about three to four hours and the scent of the plant flows through the whole house. I love it when intense odors waft throughout the house when I am boiling a dyestuff, and rhubarb gives off a very special fragrance.

After boiling, strain the root pieces and let the dye bath cool. Carefully warm the mordanted wool in the dye bath and simmer it for about an hour. While wool assumes a slightly reddish hue during the first dyeing, silk becomes green-yellow. When post-mordanting with iron water, you can create a variety of green nuances on wool as well as on silk. I don't use the leaves for dyeing myself, but they also work well for dyeing, and again call for their own experiments. The fresh leaves can also be used for dyeing, for which you may use at least three times as many leaves as dyeing fabric.

Rhubarb leaves, flowers, and roots.

LARKSPUR OR DELPHINIUM

(*DELPHINIUM* SPP.)

DYE SAMPLES

Wool

cold mordant

cold mordant
iron water

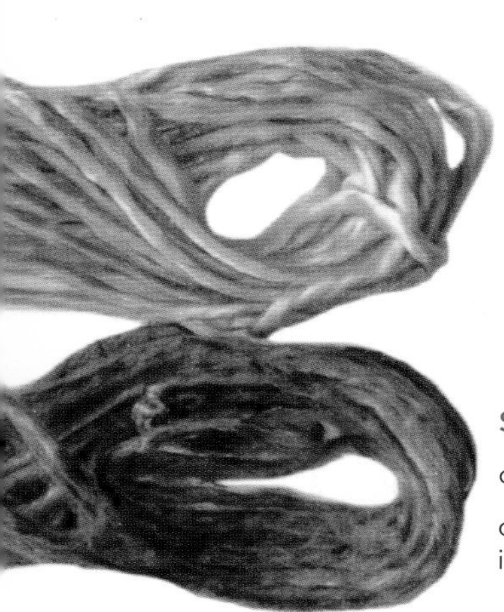

Silk

cold mordant

cold mordant
iron water

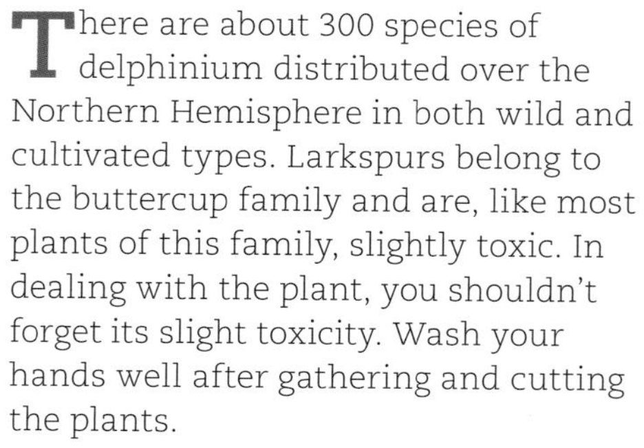

There are about 300 species of delphinium distributed over the Northern Hemisphere in both wild and cultivated types. Larkspurs belong to the buttercup family and are, like most plants of this family, slightly toxic. In dealing with the plant, you shouldn't forget its slight toxicity. Wash your hands well after gathering and cutting the plants.

You can dye well using all kinds of the conventional larkspur, the delphinium. Delphinine is, incidentally, the dyestuff responsible for the blue-violet flowers that you collect for dyeing!

Larkspur.

Dried flowers in the dye bath.

DYE RECIPE

For dyeing with larkspur, use one or two times as much of the dried blue flowers as of dyeing fabric and soak overnight. The flowers are then boiled for one hour and strained out. Add the pre-mordanted dyeing fabric to the cooled dye bath, heat carefully, and keep simmering for an hour. Then let the dyeing fabric cool in the dye bath until it has taken on the desired color.

On wool, larkspur creates a very attractive, tender green and on silk a bright lime green. For post-mordanting with iron water, add one small shot of iron water to the dye bath, add the dyeing fabric, and wait for the desired color result. On wool, a bright olive hue soon appears; on silk, the post-mordanting comes out considerably darker and you can obtain rich, dark-green shades. As is the case for all post-mordanting, you should keep a good watch on the dye bath, because your color results can change within a matter of minutes, and if you leave the pot alone, you may have missed getting the right result!

Wool and silk dyed with larkspur.

HOLLYHOCK

(ALCEA ROSEA)

DYE SAMPLES

Wool

1st dyeing
cold mordant

2nd dyeing
cold mordant

Silk

1st dyeing
cold mordant

2nd dyeing
cold mordant

Silk Overdyeing

Reseda

Marigold

Matcha tea

The hollyhock, also called the althea rose, blooms in the garden in various shades of red. The plant originally comes from southern Europe and the Middle East, but it has long since taken its place in the gardens of Central Europe, both as a healing and a dye plant. The red flowers were used for dyeing wine, liqueurs, and foodstuffs; the red-black flowers are suitable for dyeing textiles. The plant, growing up to 6½ feet high and with strong stems, is a perennial and loves very dry locations. Once the hollyhock has become established in a garden, it propagates itself, and like the reseda, it grows out of every chink in a wall and in every corner, however small.

The dried flowers can be used in tea mixtures. They display similar healing properties to the hibiscus, and this is probably where the colloquial term "farmer's hibiscus" comes from. Hollyhock tea is used for coughs, diarrhea, and stomachache. To make it, add two teaspoons of dried flowers to a cup of boiling water, let steep for ten minutes, and strain out the flowers. For acute cough or acute stomachache, drinking three cups of the tea per day is recommended.

Hollyhock, ready to prepare.

Silk dyed with hollyhock.

Hollyhock creates a pretty gray-green especially on silk, and in the second dyeing the green gives way to gray and you may produce a strong steel gray. The gray-green from the first dyeing works well for overdyeing yellow and creates different shades of green.

To do this, put the textiles dyed with reseda, Mexican marigold, or matcha tea in the dye bath and heat it carefully again, as for normal dyeing. As is the case for all overdyeing, however, it is important to stay by the pot and to check for when the desired shade is reached. Specifying a time would be purely speculative here, since the color result depends heavily on the color from the pre-dyeing. Overdyeing demands that you concentrate on what you are doing and take the dyeing fabric out of the pot immediately once the color is right!

DYE RECIPE

I collect hollyhock in my dyer's garden throughout the entire summer. Whenever flowers fall off, I collect them and lay them out to dry. In this way, even with just a few plants, by winter you will have gathered the quantity you need for a couple of dyeings. If you don't have quite enough, you can also buy hollyhock in a specialty dye store.

For dyeing, I soak the same amount of flowers as dyeing fabric for a few hours, boil this for one hour, and then strain them out. For dyeing, the pre-mordanted dyeing fabric is slowly heated and boiled gently for one hour.

Wonderful gray-green tones from hollyhock.

DYE SAMPLES

Wool

1st dyeing
cold mordant

2nd dyeing
cold mordant

1st dyeing
cold mordant
iron water

Silk

1st dyeing
cold mordant

2nd dyeing
cold mordant

1st dyeing
cold mordant
iron water

MEXICAN AND FRENCH MARIGOLD

(*TAGETES ERECTA* AND *TAGETES PATULA*)

The marigold's bright yellow-and-orange blossoms are widely distributed in Central Europe. This plant originally comes from Mexico. Both the Mexican marigold (*Tagetes erecta*) and the French marigold (*Tagetes patula*) are suitable for dyeing. In their original form, marigolds have a characteristically intense scent, and therefore the plant can be used as a natural protection against pests such as the whitefly. The scent, which many people consider unpleasant, is also repellent to the whitefly, a kind of scale insect, which particularly likes greenhouses and is very hard to get rid of. This wonderful property of insect repellent odor has been bred out of many new marigold varieties. The marigold is also used commercially for producing the dyestuff lutein, which is used in the food industry. Lutein is also used as a feed additive, for example, to get yellower egg yolks.

DYE RECIPE

Soak double the amount of dyeing fabric as of fresh, or the same amount of dried marigold, for several hours and then boil this for at least one hour. Let the dye bath cool, then strain out the blossoms, put them in a dye bag, and immerse this in the dye bath with the pre-mordanted dyeing fabric. This is heated again slowly, and the pre-mordanted dyeing

Marigolds in full bloom.

Wool and silk dyed with marigold.

fabric is simmered for up to one hour. The first dyeing with marigold yields a luminous egg yolk yellow, either on wool or on silk; the second dyeing becomes correspondingly lighter. For the second dyeing, take the flowers in the dye bag out of the dye bath. The dyeing fabric can again be left in the dye bath as desired for the second dyeing, including overnight.

This also requires using your own creativity when it comes to post-mordanting with iron water. Add a shot of iron water to the cooled dye bath and then add the dyeing fabric. Depending on what you want to create, you can now dye various shades of green, and depending on the material and how long the post-mordanting lasts, you can get lime green to olive green tones. You of course can also warm the post-mordant again, to intensify the colors. There are no limits to the dyer's joy in experimenting.

Dried marigold blossoms.

AMERICAN POKEWEED, BILBERRY, ELDERBERRY

(PHYTOLACCA AMERICANA), (VACCINIUM MYRTILLUS), (SAMBUCUS NIGRA)

The American pokeweed is one of the classic dye plants, even though during the Middle Ages the plant's berries were mostly used for dyeing foodstuffs and not necessarily for textiles. There was a tradition of dyeing skins and leather with the pokeweed berry, originally from Mexico, among the native inhabitants of North America, but not in Europe. In the Central European region, the pokeweed berry was mainly used for coloring wine, although this was not permitted in some areas. The Ulm wine ordinance of 1499, for example, explicitly forbids dyeing wine with pokeweed berries.[6] The reason for this is that all parts of the plant are classified as toxic. As is so often the case with poisonous plants, people also knew about its healing effect, if kept to the correct dose. Thus the pokeweed berry has a long tradition as a rheumatism remedy and is still used today in homeopathy. The remedy is supposed to help against rheumatism, inflammations, and the flu.

Pokeweed berries come up repeatedly in the literature on dyeing—like many other berries—and thus should not be omitted here. As already described, berries don't offer any particularly colorfast dye on textiles since the color is not lightfast in the long term and soon fades. Also, the amounts required don't speak well for the plants—for the pokeweed berry, it is reported in the literature that you need six times the quantity of berries, calculated in proportion to the dyeing fabric. For 1 kg of wool, this would be 6 kg of berries—hardly an amount that can be harvested from the bush in your own garden.

The same is true for elderberries and bilberries—although the coloring results are very intense at first, they are not very colorfast. With five times as many elderberries in proportion to dyeing fabric, you can produce a dark violet, and with this quantity of bilberries, a blue-violet. Bilberries were already in use in the Middle Ages as dyestuffs, but since they're not lightfast, these berries are not recommended as a dyestuff. In practice, I create all my blue tones with indigo and violet tones by overdyeing cochineal with indigo.

In view of the amount of bilberries or elderberries you need for dyeing, you should therefore consider whether it isn't a better idea to store the berries in jam jars or process them some other way. The flowers of the elderberry bush can also be dried and added to tea mixtures during the winter. They aid the body by their fever-sinking, sweat-inducing, and anti-inflammatory effects in case of an influenza infection. Elderberries taste wonderful

Pokeweed in flower.

6. Cf. Nübling, p. 17 ff.

when cooked together with pieces of apple and some sugar to make a compote. When used externally, crushed elderberries reveal a healing effect when used as a poultice for burns, sunburn, insect bites, headaches, or inflammations.

Bilberries taste delicious in cakes or marmalade (jam); you can make a healing tea from the leaves that works well against coughs or to alleviate mild type 2 diabetes. Bilberry leaves should be used in tea mixtures so that there is no risk of an overdose. The glycosides contained in the leaves can lead to a mild poisoning if they are taken over too long a period and the dose is too high.

DYER'S WOAD

(ISATIS TINCTORIA)

As already described in detail in the historical section, woad was one of the most important dye plants in Central European culture, before it was displaced by indigo. The dyestuff in the two plants is the same, namely indican, although both plants are assigned to different families. Indican is converted into indoxyl, a yellow dyestuff, through fermentation, and the blue indigo dyestuff can only be dissolved out through oxidation. So much for theory—in practice, however, the dye has been extracted in various ways.

While the indigo plant was cut up, layered in a large basin, and covered with water until the colorless indican turned into indoxyl through natural fermentation, the woad plants were crushed into woad pulp in the so-called woad mills and then formed into balls. These were sold to the dyers, and only then was the fermentation process started by wetting the balls of woad with water and urine. There are also records of the crushed plants being stored in sheds (barns), where a first fermentation process was started before the woad pulp was formed into balls. Basically, however, the dyestuff indican first breaks down into indoxyl and then this is subsequently oxidized during dyeing, that is, combined with oxygen. It is first through this process that the color becomes the typical blue as we know it.

Extracting the blue dye from the plant is a tedious and time-consuming affair. Due to the essentially lower dyestuff content in dyer's woad, it is hardly ever used for dyeing today. I personally do not dye using woad, but I always have a dyer's woad in my garden because it is a reminder of the ancestral art of dyeing.

Dyer's woad in a home garden.

DYE PLANTS
From Forest and Meadow

You don't have to cultivate many of the plants that work well for dyeing in your own garden, but rather they grow in forests and meadows. Here, every motivated dyer should take care when they are out gathering not to trample down the meadow when harvesting dye herbs on farmer's fields and to watch out for other plants, brooding birds, and small animals in the woods. When collecting plants in alpine meadows, be careful that you don't harvest any protected plants, that you don't destroy the sensitive alpine soil, and above all, that you don't disturb any grazing animals.

We will describe only a fraction of the countless wild plants used for dyeing in detail in this chapter, namely those plants with which I have a great deal of experience, and which I also regularly use. Plants that grow wild in our latitudes mainly dye in yellow tones, therefore most dyers in the course of their dyeing life choose some favorite plants that are abundant in their vicinity. Even though these plants can be easily dried and stored, it is preferable when using plants from the forest and meadow to turn to using fresh dyestuff instead. Over time, you acquire a good eye for the amount of plant you need to yield a dyeing; pick these and start dyeing right away.

Nettle, a healing companion that you can also use for dyeing.

With nettles for example, which grow abundantly in many places, you can dye a bright yellow, which goes towards a brownish tinge. In addition, the tea made from the nettle leaves is very healthy, and is supposed to have a blood purifying effect, and contains a lot of iron. Especially people who suffer from iron deficiency should make a practice of drinking nettle tea for two weeks' time every now and then. You should always drink the tea as a cure rather than on a regular basis.

Fern, also called male fern, grows in forests and at forest edges and dyes dark yellow; here also the fresh, whole plants are used, as with the nettle. The mugwort, which isn't related to the ferns, grows on more barren meadow edges, often along roads and on uncultivated land. You can also dye dark yellow using this plant.

When hiking through the woods and meadows, you often pass by wild raspberries and blackberries, and the leaves of these plants can also be used to produce various yellow tones. It is best to harvest these after the fruit is ripe, since before that, you might pick off the flowers and green fruit, which would be a pity!

Goldenrod at the edge of a path.

You can obtain a rich green-yellow, for example, by using wild cow parsley, a plant that grows on well-fertilized meadows. Dock and inula also dye yellow tones; the former is frequently found in pastures and is avoided by animals. Inula, on the other hand, is a wild plant that propagates on meadow edges and barren places and often appears as a volunteer in your garden. When dyeing with these plants, you can use the whole plant.

Far left: Horsetail can also be used for dyeing.

Left: Dye plants can be found in fields among the flowers and grasses.

HORSETAIL

(*EQUISETUM ARVENSE*)

DYE SAMPLES

Wool

1st dyeing
cold mordant

2nd dyeing
cold mordant

1st dyeing
cold mordant/
iron water

2nd dyeing
cold mordant/
iron water

Silk

1st dyeing
cold mordant

2nd dyeing
cold mordant

1st dyeing
cold mordant/
iron water

2nd dyeing
cold mordant/
iron water

The horsetail is a native plant and belongs to the group of fern-like plants. The plant thrives in Central Europe at altitudes up to 7,800 feet, preferably in uncultivated areas, on path and forest edges, as well as in damp meadows.

The plant, also called common horsetail, is a native medicinal plant and well-known for its anti-inflammatory properties. Tea made from horsetail is used against blistering. The plant contains silicic acid as well as various minerals and also protects the human body against skin and mucous membrane diseases. Due to its high proportion of potassium, the plant is recommended for stimulating kidney activity. To enhance cognitive performance, you should drink a cup of horsetail tea daily.

In addition to the horsetail's healing and health-promoting effects, it is an excellent dye plant for creating yellow and gray tones. To obtain these, gather the green, fresh stems in the summer. You can dye using both fresh and dried pieces of the plant; the fresh shoots create a very bright greenish yellow, and the dried plant pieces a very warm yellow.

Horsetail.

DYE RECIPE

Horsetail grows abundantly behind my house on the edge of the woods, so I prefer to dye fresh and as needed. To get the horsetail to completely release its full effect, cut the plant into small pieces and then boil these over a low flame for three hours. The water should not be allowed to boil away, but be kept just at the boil. Then let the broth cool down and you can take out the herb by hand without any problem. Of course, you can also pour it out through a sieve.

When you dye using horsetail, you again become very aware of the difference between using fresh and dried dyestuff. The horsetail that grows by my house produces bright beige tones. When I once had to order some from the supplier during the winter, I was astonished when the dried dyestuff yielded wonderfully beautiful yellow tones. Both times, the pre-mordanted dyeing fabric had been boiled for about an hour in the dye bath. To obtain gray tones, which I make mainly using horsetail, I add a shot of iron water to the dye bath after removing dyeing fabric, and re-immerse the dyeing fabric. The palette of gray tones will also now vary, depending on what beige tone you obtained from the dyeing. Thus you need to practice here if you want to obtain a very specific gray tone. In addition, not only is it important for the dyeing result, whether you use a fresh or dried dyestuff, but where the plant grew and the weather conditions also matter. There will always be surprises when you dye with plants!

Dried horsetail.

Horsetail creates wonderful shades of gray on silk.

GOLDENROD

(SOLIDAGO VIRGAUREA)

DYE SAMPLES

Wool

1st dyeing cold mordant

2nd dyeing cold mordant

1st dyeing cold mordant/ iron water

Silk

1st dyeing cold mordant

2nd dyeing cold mordant

1st dyeing cold mordant/ iron water

The unpretentious goldenrod grows almost everywhere in our latitudes: on the edges of farmland, in forest glades, in fields and meadows. One of the folk names for goldenrod is "woundwort" or "gold woundwort," which indicates the anti-inflammatory effect of the plant. From a healer's point of view, the goldenrod is quite simply the kidney plant, since it has a very diuretic effect and stimulates the metabolism. A tincture of goldenrod is a very simple remedy to support kidney function: Add 6 g of goldenrod flowers to 100 ml of white wine, let it steep for a week, and filter out the flowers. Drink a schnapps glass full at meals! Saint Hildegard of Bingen always swore by goldenrod herb, which was boiled and the broth used to treat throat inflammations. It worked wonders on open sores in the mouth.

The goldenrod is a gratifying plant for dyeing. I harvest goldenrod as a whole plant, when the flowers have just freshly blossomed and hang them up to dry. When the plants are well dried, cut them into small pieces and

A feast for the bees: goldenrod in high summer.

store the plant pieces in a paper bag or carton.

DYE RECIPE

For dyeing, use the same amount of goldenrod as dyeing fabric. Soak the plant pieces overnight and then boil for two hours. Then strain the dyestuff out; it isn't added back into the dye bath. Slowly bring the pre-mordanted dyeing fabric to a boil in the dye bath and boil for about an hour; it can be removed as soon as it has reached the desired color. Goldenrod dyes sun yellow and you can use the dye bath again for a second and third dyeing. These dyeings turn out correspondingly lighter.

If you want to obtain yellow-green (on wool) or dark green (on silk), then post-mordant with iron water. To do this, take the dyeing fabric out of the dye bath, add a shot of iron water, and return the dyeing fabric to the dye bath. Leave it there until the desired green tone is reached. It is particularly striking about goldenrod that the results of post-mordanting are much more intense on silk than on wool.

Goldenrod in a field.

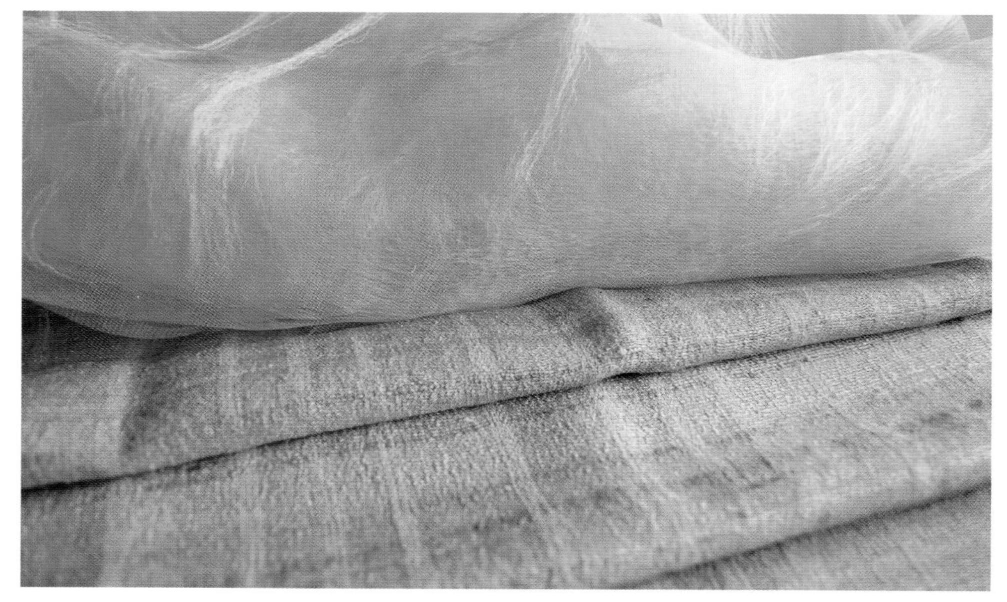

Silk dyed with goldenrod.

ST. JOHN'S WORT

(HYPERICUM PERFORATUM)

The delicate St. John's wort flowers shine like small suns in June. While earlier on, St. John's wort grew in meadows and on field edges, today it is only rarely found there, and this healing herb has retreated to the garden. It is therefore all the more important to grow St. John's wort in your own herb or dyer's garden! In addition to its healing effect—St John's wort is added to self-harvested herbal tea and is used to improve your mood on dark winter days—St. John's wort is also an important food for bees.

Even back as far as Paracelsus, St. John's wort was recommended because of its blood-purifying and wound-healing effects. Many people still know red St. John's Wort oil from their childhood, when it was applied generously on all injuries. It is simple to make St. John's wort oil if you exercise some patience. Gather the fresh flowers and buds, put them in a glass jar, and pour on some good olive oil. Different quantities for this are given in the literature, but you can't go wrong with two handfuls of

St. John's wort in full bloom.

DYE SAMPLES

Wool

1st dyeing
cold mordant

2nd dyeing
cold mordant

1st dyeing
cold mordant/
iron water

Silk

1st dyeing
cold mordant

2nd dyeing
cold mordant

1st dyeing
cold mordant/
iron water

Wool and silk dyed with St. John's wort.

blossoms to 1 liter (0.26 gallons) of oil. Now place the glass jar in the sun and shake it daily for about eight weeks until the oil is dyed dark red. Then strain out the flowers and store the oil in a cool place. It is used externally for rubbing on tight muscles or sunburn or internally against nervousness or menopause symptoms. The oil should be taken lukewarm, two teaspoons a day. Taking St. John's wort both externally and internally makes the skin more light-sensitive; therefore, you should be careful about being in direct sunlight!

St. John's wort dyes yellow-green to olive green, whether used fresh or dry. For drying, harvest the St. John's wort before the flowers turn brown. Cut down the whole plant and leave it to dry well with the flowers hanging downwards. After that, cut the herbs into small pieces and store the plant pieces in a paper bag or carton.

DYE RECIPE

For dyeing, I use the same amount of dried dyestuff as dyeing fabric or double the amount of fresh herb. Soak the fresh herb for two to three hours and the dried ones overnight, then boil for two more hours and strain it out. As described in the basic recipe, bring the pre-mordanted dyeing fabric slowly to a boil and boil it for about two hours in the dye bath. For post-mordanting with iron water, take the dyeing fabric out of the dye bath, add a shot of iron water, and return the dyeing fabric to the dye bath. Now you can observe how the materials become ever darker; silk takes on the color from St. John's wort very well and will dye even a dark olive green, while wool always remains somewhat lighter.

The dried herb is used for dyeing.

MEADOWSWEET

(*FILIPENDULA ULMARIA*)

DYE SAMPLES

Wool

1st dyeing
cold mordant

1st dyeing
cold mordant
iron water

2nd dyeing
cold mordant

2nd dyeing
cold mordant/
potash

Silk

1st dyeing
cold mordant

1st dyeing
cold mordant
iron water

2nd dyeing
cold mordant

2nd dyeing
cold mordant/
potash

The true meadowsweet grows especially well in the open countryside along brook and river banks and in damp meadows, but can also be cultivated in the garden. Here it is very happy in a place by the garden pond. According to legend, the plant, also called wild lilac or mead wort, was said to have first been sown by Mary, enshrined in popular faith as the protector of medicinal plants. The plant contains salicylic acid, the agent from which aspirin was synthesized, and is used as a tea for headaches, colds, flu, or rheumatism. To make the tea, pour a quarter-liter (0.52 pints) of boiling water over the fresh or dried herbs and let steep for ten minutes.

True meadowsweet is native to Central Europe and parts of Asia and was used earlier to sweeten meads and wines. Once you have smelled the intense, honey-like fragrance of the flowers, you can well understand why. In July, true meadowsweet is in full bloom in the places where it is still allowed to spread, and beguiles you with its sweet summer fragrance.

Meadowsweet.

Silk dyed with meadowsweet.

DYE RECIPE

Meadowsweet is harvested in July; then you should dry the plants and store them dry in a cardboard box. For dyeing, cut the plants up, use the same amount as the dyeing fabric, and soak overnight. It is also possible to dye using the fresh plant, and for this use double the amount by weight of the plant in proportion to the dyeing fabric. Bring this to a boil without soaking. After simmering the dye bath for two hours, strain out the dyestuff and immerse the pre-mordanted dyeing fabric in the dye bath. Heat this slowly and simmer it for a half hour to an hour.

With meadowsweet, it makes a big difference whether you use the whole plant or only the flowers. When the whole plant is used, you obtain a rich brown-yellow, which is reminiscent of late summer, and if you use the flowers, a lighter yellow emerges. Textiles dyed both using the whole plant as well as just the flowers can be post-mordanted, both with water and with potash.

Post-mordanting with iron water yields green tones while post-mordanting with potash intensifies the yellow to a magnificent sun yellow. Both post-mordanting processes are similar: Add either a shot of iron water or a teaspoon of potash to the dye bath after the dyeing fabric is removed, then return the dyeing fabric to the dye bath. Then wait until it takes on the desired color. I find it remarkable that with potash I can also obtain a very intense yellow from the second dyeing.

YARROW

(ACHILLEA MILLEFOLIUM)

DYE SAMPLES

Wool

cold mordant

cold mordant
iron water

Silk

cold mordant

cold mordant
iron water

Yarrow, which is native to Europe, grows in lush meadows that are not too heavily fertilized. The yarrow, unfortunately, has long since disappeared from the scene in industrially farmed land. A good place for finding yarrow is farms that keep horses, since the meadows on these farms are mostly used as paddocks and not to raise large grass and hay yields. We can also still find yarrow in alpine meadows. Both white and red yarrow can both be used for dyeing, and if, when gathering yarrow, you also put aside some blossoms to dry for tea, you will have one of the best medicinal herbs of our latitude available for your use. Yarrow tea works against digestive symptoms and all so-called women's symptoms, from menstrual pain to menopause symptoms. If you want to mix a helpful tea for menopause, collect equal amounts of yarrow, raspberry leaves, and lady's mantle leaves before the plants bloom, and the blossoms of red clover. Dry the herbs well and then enjoy them daily in a tea.

Yarrow is also considered a magically protective herb and was used as one of the plants to make up herb wreaths against evil spells. Perhaps it also exercises this protection in herbal schnapps? The recipe is very simple: Gather the pink-colored yarrow blooms, put them in a glass jar, and fill it up with as much grain alcohol or schnapps until the flowers are entirely covered. Then let the mixture age for two weeks

Yarrow galore.

Yarrow-dyed wool.

in a cool place, strain out the flowers, and bottle the yarrow schnapps. The longer the schnapps is allowed to age, the more its taste emerges!

DYE RECIPE

The yarrow is a very gratifying dye plant. If you have a meadow nearby where yarrow grows, you will quickly be able to collect enough plants for a dyeing. You can dye using both fresh and dried plants; here again the quantity is double by weight of fresh dyestuff or the simple weight of dried dyestuff in proportion to the dyeing fabric. You can dye right away with the fresh plants, but the dried plants are soaked overnight. The flowers are then boiled for one hour and finally strained out. Put the pre-mordanted dyeing fabric into the dye bath and slowly heat it again. After an hour of simmering, you get a rich yellow on wool and a more delicate yellow-green on silk. Post-mordanting with iron water yields a wide variety of green shadings on wool as well as on silk. To do this, take the dyeing fabric out of the dye bath and wring it out, add a shot of iron water to the dye bath, and again immerse the dyeing fabric.

As with most post-mordanting, stay close to the pot and take the materials out when they have taken on the desired hue.

Fresh yarrow flowers.

DYE FROM TREES, ROOTS, AND BARK

Oak, Ash, and Alder Buckthorn Bark

The roots and bark contain the plant's strongest dyeing power. Roots and root pieces are available from specialty dyeing suppliers, as are bark pieces. However, you can also collect and process the bark yourself when a tree is cut down by asking for any branches that are no longer needed. In this case I particularly recommend the thin branches, since they can be used in their entirety and you don't have to peel them besides.

I have experienced this with fruit woods, but you can certainly also ask for the branches if a tree is felled in the woods near where you live.

Anyone who works with purchased wood gets it already chopped small; when you are gathering it yourself, take the branches home, cut them up with tree shears, and let the pieces dry.

The ash grows on the forest edges.

Many woods from the forest are suitable for dyeing.

As with rhubarb root, it is a good idea to use a dehydrator if you have one. Although branches always look very dry, if they aren't completely rotten, they are moist inside and there is a risk that you won't leave them to dry out long enough. You may be patient with branch pieces! Rotten branches, incidentally, aren't good for dyeing.

With native woods, you can dye to get a palette between brown and beige, which, with post-mordanting, yield gray hues in all shadings. When dyeing with wood, each dyer will soon get their own favorite, because it is of course more fun to be able to use the woods in your own neighborhood than to just buy it. Thus, in this chapter I describe those trees that I use for dyeing myself and also with which I was able to gain a lot of experience. Don't just use the bark or wood from all these trees; the leaves also work well for dyeing. This will be precisely described for the respective trees.

Among the trees not mentioned here are, for example, the horse chestnut tree, which can be used like the walnut. You can use the green shells of the nuts for dyeing and obtain green-yellow and green; you can use a classic dyeing process as well as cold dyeing. This creates a lighter brown result than the walnut. The wood of the alder gives similar coloring results as the ash or oak, which are described in detail in this chapter.

An old nut tree—a feast for every dyeing studio.

The apple tree's bark also dyes.

COMMON BIRCH, WHITE BIRCH, SILVER BIRCH

(BETULA PENDULA)

DYE SAMPLES

Wool

1st dyeing

2nd dyeing

1st dyeing iron water

2nd dyeing iron water

Silk

1st dyeing

2nd dyeing

1st dyeing iron water

2nd dyeing iron water

The birch is a wonderful healing and dyeing plant. The birch is the symbol of spring, of fertility and a new beginning, and textiles dyed with birch radiate this vitality. Because of its ability to flush out water accumulated in the human body and prevent kidney disease, the birch is called the "kidney tree." Birch leaf tea is also supposed to have miraculous effects against rheumatic diseases. Fresh birch leaves are used for a kidney-strengthening full bath. To do this, pour about 10 liters (2.64 gallons) of boiling water over two handfuls of fresh birch leaves and pour this in the bathtub. For birch leaf tea, as for the full bath, pick the young leaves as long as they are still very small and somewhat sticky.

These fresh leaves can also be used for dyeing; they should be picked in May, and then in June you should pick your yearly stock and then dry it. Store the dried leaves in a carton or a paper bag. The dried leaves can be also used as tea just as for dyeing. The dried birch leaves yield a wonderful gold-yellow and the fresh leaves make the color slightly greener and brighter.

DYE RECIPE

For dyeing with birch leaves, soak them overnight and boil for one hour before the dyeing process. For fresh birch leaves, use double the quantity of leaves in proportion to material; for dried leaves the same quantity by weight as material is used. With birch, you can do a so-called direct dyeing, which means that the fabric doesn't have to be pre-mordanted, but rather you add the mordant right to the dye bath. To do this, after the boiling process, put the leaves in a dye bag, close it, and put it back in the dye bath. Now add the calculated amount of alum to the dye bath, meaning that the dyeing fabric should be treated with 10% alum by weight. Thus for 1 kg of dyeing fabric, dissolve 100 g of alum in warm water and add this to the dye bath.

Birch at the roadside.

Now add the dyeing fabric, heat the dye bath again slowly, and simmer it for half an hour to an hour. Birch has an abundance of dye, so it is possible to do several dyeings. While the dyeing fabric takes on a radiant spring yellow in the first dyeing, the further dyeings become ever lighter, up to a pale but still bright yellow. Post-mordanting with iron water gives fresh green tones. To do this, take the dyeing fabric out of the dye bath and add a shot of iron water, put the dyeing fabric back in the dye bath, and leave it there until the desired green tone is obtained. The wonderful yellow tones from dyeing with birch, by the way, also work excellently for overdyeing, for example with indigo or madder.

Wool yarn and cloth.

Wool dyed with dried birch leaves.

ENGLISH OAK

(QUERCUS ROBUR)

DYE SAMPLES

Wool

without mordant

cold mordant

without mordant iron water

cold mordant iron water

Silk

without mordant

cold mordant

without mordant iron water

cold mordant iron water

Since antiquity, the oak has been highly revered in Central Europe, on the one hand due to the magical stories surrounding this tree, and on the other because the oak was a food source. As far as magic is concerned, oaks are supposedly struck by lightning more often than other trees because they attract the lightning. That is why the oak was known as the lightning tree and dedicated to the supreme gods. Among the ancient Greeks this was Zeus, among the Romans Jupiter, and among the Germans Thor.

As a source of food, we can say that a flour was made from acorns, the fruit of the oak, which was a valuable source of energy. Still today, acorns are processed as food: If you roast acorns in the oven like chestnuts, then you can grind them in a coffee grinder; this way, you can make the famous acorn coffee. To brew it, just pour water over the powder like for making malt coffee. According to folk medicine records, this acorn coffee is supposed to help against "exhaustion and depression" as well as against stomach and digestive symptoms. If you want to make this acorn coffee yourself, you have to pick the ripe acorns from the tree and not off the ground. The acorns lying on the ground are usually already infested with worms.

Tea made from oak bark is recommended against diarrhea, and

Fresh oak leaves in late spring.

Wool in all shades, dyed with oak.

full baths in oak bark are said to help against muscle rheumatism. The oak bark is also used for dyeing, and, depending on how it is prepared, you obtain results ranging from light brown to gray. You can get oak bark both from your dye supplier and in every pharmacy.

DYE RECIPE

For dyeing with oak bark, use the same amount by weight of dried bark pieces as dyeing fabric and soak these overnight. The next day, heat the dye bath and boil it for two or three hours, then strain out the pieces of bark into a dye bag and return it to the dye bath in the bag. Now add the dyeing fabric. The special feature of dyeing with oak bark is that you don't have to use mordanted material. Oak bark contains tannins, which are normally used for pre-mordanting! The results vary—while un-mordanted material is skin-colored to beige-brown, on pre-mordanted material oak bark produces light brown to coffee brown tones. To do this, the material must be simmered for one or two hours in the dye bath. Here you can also post-mordant with iron water; take the dyeing fabric out of the dye bath, then add a shot of iron water to the dye bath and then re-immerse the material that you want to post-mordant. Both pre-mordanted and non-mordanted material now takes on a strong gray tone, which will be slightly lighter on non-mordanted material. As always, however, with post-mordanting, this hue depends very much on how long the post-mordanting process is.

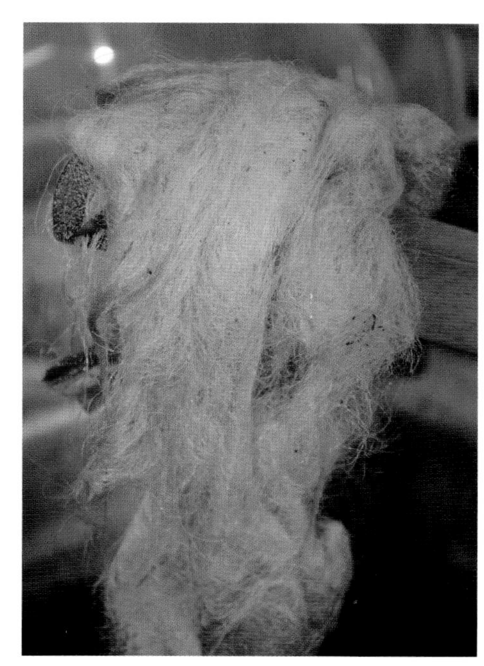
Wool, dyed with oak, is removed from the dye bath.

OAK GALL (OAK APPLE)

DYE SAMPLES

Wool

1st dyeing

1st dyeing cold mordant iron water

1st dyeing iron water

2nd dyeing iron water

Silk

1st dyeing

1st dyeing cold mordant iron water

1st dyeing iron water

2nd dyeing iron water

The tannin that you use as a pre-mordant comes from the so-called oak apple or oak gall. These round objects grow on branches and on the underside of oak leaves after the oak gall wasp damages the bark or leaves and then lays its eggs in that place. The plant's defense reaction causes these apple-shaped oak galls to form, small balls in which the larva of the oak gall wasp grow. When the larvae leave the oak apple in spring, then these can be used to extract tannin. Earlier on, due to their tannin content, oak apples were used for making black ink or tanning leather.

Since collecting oak galls is rather tedious, especially from tall, old trees, it is a good idea to get your tannins from oak galls from a dyeing accessories supplier. There you can also get the oak galls themselves. Oak galls are not only suitable as pre-mordanting but you can also use them directly for dyeing. Normally, you use the commercially available powder from oak galls, but if you collect whole oak galls or buy them, they must be ground into a powder.

DYE RECIPE

As with all substances containing tannin, it is possible to dye with oak galls without any pre-mordanting, but you get significantly lighter results than if you dye using pre-mordanted material. For

Oak galls on dyed silk cloth.

this reason, I use also pre-mordanted material when dyeing with oak galls. For dyeing, take one-tenth oak gall powder by weight in proportion to the material and stir it into the dye bath. Thus, for 100 g of wool or silk, use 10 g of oak gall powder; for 1 kg of material, 100 g of powder is needed. You can also use tartar, and if you do this, use half of the amount for oak galls. Tartar intensifies the red share of a color, and if dyeing with oak gall, you obtain a gray hue with something of a red tinge by adding tartar.

If you want to work with tarter, stir the tartar powder into the dye bath, then let this come to a boil and boil it for half an hour. The dyeing fabric is then added to the cooled dye bath and, as in a conventional dyeing process, is carefully heated to the boiling point. You can take out the dyeing fabric at any time, when you obtain the desired hue. You can make many post-dyeings with oak galls; every time, the results are a little lighter and thus you can create a whole range of different gray tones. The first dyeing can be so dark that you have almost obtained a black!

Since I personally find the gray tones of the horsetail more appealing, I usually use it to dye gray. Dyeing with oak gall powder is an excellent alternative, if you need gray material quickly or have run out of your supply of horsetail.

Silk cloths (top) and yarn (below) dyed with oak gall.

COMMON ASH

(FRAXINUS EXCELSIOR)

DYE SAMPLES

Wool

cold mordant

without mordant

cold mordant iron water

without mordant iron water

Silk

cold mordant

without mordant

cold mordant iron water

without mordant iron water

The ash is a tree native to Europe, which was credited with magical powers by the time of the Celts. The tree stands for the power of water, which may be due to the fact that springs of water are often found near stands of ash trees. No wonder, since the ash grows particularly well in damp forests and meadows. In Norse mythology, the ash was designated as a world tree, at whose roots the Norns, the goddesses of destiny, sit and spin both the links between the worlds and the fate of human beings.

In addition to all the mythological significance of the ash, this tree has always been used for remedies. Tea made from the bark or from the leaves has a diuretic and laxative effect and can thus be used for constipation. An infusion of the leaves may be administered against rheumatism and gout by applying the leaves.

Ash bark is available both as a dyestuff and as tea.

Wool dyed with ash, spun and unspun.

DYE RECIPE

The ash bark is used for dyeing; this can be bought already crushed from a dyeing accessories retailer or in the pharmacy. Of course, you can also use small branches for dyeing by cutting them into small pieces with tree shears, letting them dry well, and then storing them in a paper bag or carton. As with all woods, you should make sure that you let the pieces of wood dry thoroughly, since they contain a lot of moisture. You should not use rotten wood.

For dyeing, use the same quantity by weight of ash wood as dyeing fabric. Soak the pieces at least overnight or up to two days; the next day, bring the dye bath to a boil and boil for two or three hours. Since ash wood also contains tannins, you can dye un-mordanted fabric, but pre-mordanted fabric produces more intense colors. After boiling, put the wood pieces into a dye bag and put it back into the dye bath. Now add the dyeing fabric, heat it carefully, and simmer for about an hour.

Dyeing with ash bark yields a light yellow to yellow tone, and post-mordanting produces color shadings of green and gray. This is totally a matter of whether or not you have pre-mordanted or not—cold-mordanted wool, for example, produces green tones in post-mordanting, and wool that is not mordanted a medium gray hue. Iron water can also be used for post-mordanting; add a shot after you take the dyeing fabric out of the dye bath. Now you can put any pieces that you want to post-mordant back in the dye bath and wait until you obtain the desired colors.

Silk cloth dyed with ash.

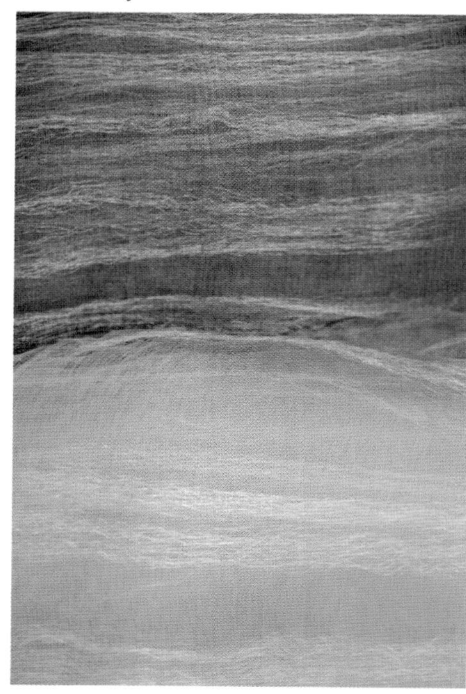

ALDER BUCKTHORN

(FRANGULA ALNUS)

DYE SAMPLES

Wool

1st dyeing

1st dyeing iron water

1st dyeing potash

Silk

1st dyeing

1st dyeing iron water

1st dyeing potash

The alder buckthorn is a common shrub species in Central Europe, and from there it was introduced to North America. It is also native to Asia and North Africa. The alder buckthorn prefers damp soils and therefore grows especially well at forest edges and in riparian forests. Alder buckthorn bark was already used as a laxative by the late Middle Ages, but you can only use bark you have collected yourself after if it has been allowed to dry for at least one year. The long-term storage lets the toxins in the bark degrade. These are also found in the leaves and berries, so neither the leaves nor the fruit of the alder buckthorn should be consumed. Even today, alder buckthorn bark is used in teas that have a laxative effect, and also has a place in homeopathy—as a remedy against constipation. The alder buckthorn is also popularly known as "powder wood" because its wood was processed into charcoal and then into gunpowder. For dyeing, it is advisable to buy the dyestuff in a specialty store.

DYE RECIPE

For dyeing, soak the alder buckthorn bark pieces for several days and then boil it for three hours and strain into a dye bag. Use the same amount of bark by weight as dyeing fabric. Re-

Alder buckthorn.

Wool dyed with alder buckthorn bark.

What is special about dyeing with alder buckthorn for me is post-mordanting with potash. To do this, take the dyeing fabric out of the dye bath and set it aside briefly. Now dissolve one tablespoon of potash per 500 g of dyeing fabric in some water and pour it into the dye bath. Put the dyeing fabric in this post-mordant and all kinds of different pink shades develop! Depending on the material and how long you post-mordant, the nuances of color can range from a delicate light pink to an intense dusky rose. It is also possible to produce two-color skeins of wool by hanging only half of the skein in the post-mordant. As with all overdyeings, the results of the dyeing harmonize very well with those of the dyeing plus post-mordanting, and with very little effort you are able knit in two colors.

immerse the dye bag together with the dyeing fabric in the dye bath and slowly bring it to a boil. After about an hour of simmering, you will get rich yellow hues on pre-mordanted material, which transform into yellow-green when you post-mordant with iron water. To do this, take the textiles out of the dye bath, add a shot of iron water to the dye bath and re-immerse the material in the dye bath. Most of the time, the results of post-mordanting are significantly more impressive on silk than on wool.

Alder buckthorn bark.

WISTERIA, CHINESE WISTERIA

(*WISTERIA* SPP.)

DYE SAMPLES

Wool

1st dyeing
cold mordant

2nd dyeing
cold mordant

1st dyeing
cold mordant
iron water

Silk

1st dyeing
cold mordant

2nd dyeing
cold mordant

1st dyeing
cold mordant
iron water

The wisteria that is so widely distributed in our gardens and also known as Chinese wisteria, is a plant genus to which some ten species are assigned. The genus is native to North America and Asia. Chinese wisteria is particularly enchanting in spring as well as in late summer with its splendid purple flowers. The plant can grow several yards high and climbs up trellises or balconies. Every garden enthusiast who has a wisteria at home also knows the problems with this enchanting growth—it often can take years until the wisteria finally emerges in its flowering splendor, but at the same time it climbs so vigorously that it must be cut back several times each year to make sure it doesn't take over the whole garden. When it finally blooms, however, it does this with an intensity that can only leave us astounded and usually does so twice a year. You can encourage the second blooming by skillfully pruning the plant after the first bloom. The wisteria can keep growing until it is up to one hundred years old!

In itself, the wisteria isn't a distinctive dye plant, but I was inspired by the wisteria that has already grown into a tree in my own garden. What decided me in my attempt to dye with wisteria leaves was my first experiment in the area of eco printing, which is also described in this book. From the results of this eco printing, I got the desire to use wisteria leaves for

In the spring the wisteria shows its first blossoms.

conventional dyeing—there are finally enough leaves available on my plant.

DYE RECIPE

For dyeing, use the fresh leaves on their stems, and use three to four times the quantity by weight in proportion to dyeing fabric. Cut the leaves up roughly and boil them in a large pot for an hour. Then strain out the plant pieces and add the pre-mordanted material to the dye bath, where you carefully bring it to a boil and keep it simmering for between a half hour and an hour. The results of the first dyeing are a very luminous sun yellow and the second dyeing yields a light, radiant yellow-green.

You can also post-mordant with wisteria dyeing; here, too, I prefer using iron water, which brings out green hues in various shadings. Take the yellow-dyed textiles out of the dye bath and add a shot of homemade iron water to it. Then re-immerse the dyeing fabric in the dye bath and wait until the desired green is reached.

Silk and wild silk dyed with wisteria.

Wisteria-dyed wild silk and a wisteria eco print.

WALNUT

(JUGLANS REGIA)

DYE SAMPLES

Wool

1st dyeing

2nd dyeing

3rd dyeing

1st dyeing iron water

2nd dyeing iron water

3rd dyeing iron water

Silk

1st dyeing

2nd dyeing

3rd dyeing

1st dyeing iron water

2nd dyeing iron water

3rd dyeing iron water

The walnut tree comes originally from southern Europe and Central Asia, but now North Americans too would find life hard to imagine without its nuts. The delicious walnut kernels contain polyunsaturated fatty acids and are very healthy, while the tree leaves are used as a blood purifier. But this isn't the only area where tea made from walnut leaves has a healing effect. Walnut leaf tea is also known in folk medicine as a remedy for rheumatism or for inflamed mucous membranes. No matter whether you are struggling against mucous membrane inflammations in the intestine, stomach, or scalp area, walnut leaf tea is a tried and true remedy!

For gourmets, we recommend you make your own nut liqueur. The recipe is extremely simple and is available in many variations. The following is one of the easiest: Chop up ten green nuts and let steep in 1 liter (0.26 gallons) of fruit schnapps for six weeks. Then boil 250 g sugar down to a syrup in 1 liter (0.26 gallons) of red wine, mix the two liquids, pour into bottles, and wait another week. Now the nut liqueur is ready to enjoy!

But back to dyeing—fresh nutshells dye remarkably well, so you should always wear gloves when processing the fresh shells!

In the autumn, the nut trees beckon with their rich harvest.

In autumn, when the nuts fall from the trees, I always drive to my "nut farmer's wife," as I secretly call her. Behind the old farmhouse on the hill stand four large, old walnut trees, which always yield a good harvest. I pick up both fresh nuts and the nutshells there and process them immediately for cold dyeing with nuts. The chapter on dyeing (see page 14) describes in detail how cold dyeing works, and the color palette ranges from dark yellow to light brown.

DYE RECIPE

Dyeing with dried walnut shells works by direct dyeing and yields colors ranging from light brown to gray-brown, depending on the material. In addition, dyeing with walnut also yields the basic color for obtaining muted colors. First you pre-mordant and then dye with walnut shells and then use a further dyeing process to overdye with indigo or madder. Pre-mordanting is necessary for overdyeing, otherwise the pre-dyed material will not take on the indigo. For dyeing with dried walnut shells, refer to the basic recipe; you should use the same amount of dyestuff as dyeing fabric. The dye samples show the results of direct dyeing; these weren't done using pre-mordant.

Dried nutshells ready for use in dyeing.

Wool and silk cloth dyed with walnuts.

Further Processing by Cold Dyeing

After cold dyeing, you can use the dye bath again! To do this, just bring the dye bath with the nutshells in it to a boil and boil it for two hours. Then strain out the nutshells and continue to dye in the remaining dye bath. The resulting dyed textiles obtain wonderful nuances of dark brown color.

Of course, you can also skip the cold dyeing process and the green nutshells can be processed as in the basic dyeing recipe. I also repeatedly make attempts to get more intense colors by leaving the dyeing fabric in the dye bath—for several

Fresh green nutshells for cold dyeing with walnut.

hours or even for several days. The results can't be described in detail here, since they vary greatly. Dyeing results depend very much on the material, the quality of the dyestuff, the time of harvesting, and in my opinion, on the quality of the day. Here it is worth your while to make your own experiments, to test and to sample. If the results are particularly beautiful, you should note down the process so that in the future you can obtain these results again.

Overdyeing Textiles Pre-dyed with Walnut

Material dyed with walnut can be overdyed beautifully with other colors, so you can complete your color palette with fine nuances. For this, it is important to note that only the beige and light brown pieces are suitable for overdyeing. In addition, the textiles that are to be overdyed must in any case be pre-mordanted. You can do countless rounds of dyeing with a walnut dye bath, but with each round the dyeing becomes lighter. I store these lightly dyed materials for overdyeing. Overdyeing is done in the same way as a normal dyeing process, but instead of untreated wool, simply use the pre-dyed material.

With walnut, material dyed beige reveals itself, when overdyed with indigo, in the nuances of petrol; if overdyed with madder, in various reddish-brown tones; and with yellow colors from dyer's chamomile, birch, or marigold, in brown-yellow in various gradations of color.

Unspun wool cold dyed with nuts.

Spun wool cold dyed with nuts.

DYEING WITH INGREDIENTS FROM THE KITCHEN

Who hasn't experienced this: a lively evening in the yard, stimulating entertainment, you're barbecuing, and the children are playing in the grass. The final results: The red wine stain on the tablecloth that can never be completely washed out, the grass stains on the children's shorts, and the mustard spot on your white blouse. You might suppose that the foodstuffs that produce such stubborn stains would also be wonderful for dyeing!

Of course you can dye using foods you use every day. Many seasonings yield wonderful colors, and tea and coffee also work well for dyeing textiles. However, it is mostly the foodstuffs that leave the most stubborn stains that are not suitable for dyeing. For example, you can easily see this for yourself with red wine—the stains change after washing or by exposure to the sun from red to purple to a pale pink; red wine is thus not a lightfast dye and fades very quickly. The same applies to mustard and pumpkin seed oil, fortunately for all those who tend to spill them.

But there are a lot of kitchen ingredients that are more lightfast than those described above, and which are therefore excellently suited for experimentation. In addition, it is a lot of fun to use something that otherwise ends up in the trash, such as carrot greens or onion skins.

Ingredients from the kitchen.

RED CABBAGE

(*BRASSICA OLERACEA* CONVAR. *CAPITATA* VAR. *RUBRA*)

DYE SAMPLES

Silk

cold mordant/ salt
iron water

cold mordant/ salt

cold mordant

alum/ tartar

Red cabbage is a typical winter vegetable and keeps very well. The different names come from the method of preparation in individual regions—the more acidic the preparation, for example by adding vinegar, the redder the vegetable appears, while the more alkaline, the bluer. This is already true while the plant is still growing; red cabbage is a so-called acid-base indicator and reacts to different soils by changing its color. The lower the pH value, the more acidic a soil is and thus the redder the plant leaves become, while the more alkaline a soil, the bluer the leaves appear.

For dyeing with red cabbage, you shouldn't use the young summer red cabbage, but rather the winter red cabbage, which is harvested in autumn. This obviously contains more dyestuff, and I obtained the illustrated dye samples by using it.

DYE RECIPE

Cut five times the amount of red cabbage in proportion to dyeing fabric into strips and layer it with the premordanted dyeing fabric in a pot. This means a layer of red cabbage, a layer of silk, and again a layer of red cabbage. Then fill the pot with water warm to the hand and let everything soak overnight. The next day, heat the dye bath carefully to a maximum of 176°F (80°C), leave it at this temperature for one to two hours,

Red cabbage.

A skein of silk dyed with red cabbage.

and finally let it cool. Then take the dyeing fabric out of the dye bath; after you let it dry it will get a rich purple hue. In my experience, red cabbage dyes silk very well, while you only get very dull gray-purple hues on wool. This is why there is no wool at all in the dye samples at left. To dye wool purple it works much better to overdye cochineal with indigo. Many dyers are engaged in the search for the perfect purple tone that you can get on wool without overdyeing, and there is still plenty of room for experimenting here. Dyeing with red cabbage, whether using a cold mordant or fabric pre-treated with alum, doesn't produce any satisfactory result on wool.

But back to the silk dyes! If you want to continue to dye the dyeing fabric used in the first process, you can also obtain blue tones. To do this, add a tablespoon of salt to the pot with the red cabbage, and the selected dyeing fabric is returned to the dye bath. There it is heated again to 176°F (80°C) and left to cool down afterwards in the dye bath. Take the dyeing fabric out when it has taken on the desired blue tone.

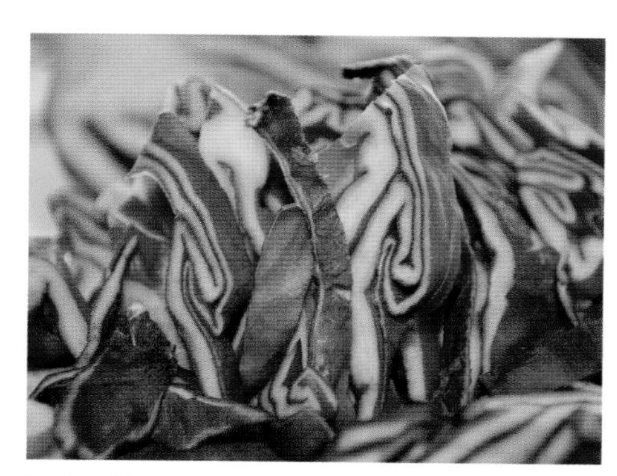

Red cabbage cut in strips for dyeing.

COFFEE

(COFFEA)

DYE SAMPLES

Wool

cold mordant

cold mordant
iron water

Silk

cold mordant

cold mordant
iron water

The coffee plant belongs to the family of bedstraw plants, the same family as madder. At present there are 124 known species of *coffea* and ten of them are currently known to be cultivated; the seeds are harvested from at least six species, then roasted and processed into coffee as we know it. There are a number of legends about the moment when man discovered coffee. One is told of the observations of a goatherd, who saw that if his animals grazed on a particular shrub, they went leaping around all night. Another legend tells of a shepherd who assumed the plant's seeds were inedible, spit them into the fire, and discovered how to roast coffee because of the resulting pleasant fragrances.

Whatever comes closest to the truth, it is proven that coffee had already arrived in Europe in the sixteenth century. By the seventeenth century the first coffee houses were opened, in Bremen (1673) and in Vienna (1685), for example. Almost all coffee species originated in Africa, but they have been cultivated in the tropical and subtropical zones all over the world and became indigenous there.

Coffee beans and ground coffee.

Wool and silk dyed with coffee.

Meanwhile, coffee is now cultivated in more than fifty countries. The coffee beans from which we extract our coffee are the seeds of the coffee plant fruit. These seeds have to be cleaned of all plant fibers and dried and then roasted, before they can be used to make the hot beverage. The varying methods of roasting allow us to produce a wide variety of coffee qualities.

Drinkers of strong coffee, as well as tea drinkers, can report that both substances dye wonderfully well. By the way, you can whiten teeth that are stained by drinking a lot of coffee and tea by brushing your teeth with turmeric powder! To do this, dip your moistened toothbrush in the powder and clean your teeth as with toothpaste. After only a few weeks, you can see a visible result.

DYE RECIPE

Coffee is found in almost every household and sometimes you find a kind that isn't particularly to your taste or was left in a corner of the cupboard and forgotten. These leftovers can be used in the dyer's studio. For dyeing, I use the same amount of coffee as dyeing fabric; if you use more coffee, the results are correspondingly darker. Boil the ground coffee for about half an hour, let it cool, and filter it. Then add the pre-mordanted dyeing fabric to the dye bath. It isn't a problem if any coffee grounds were left in it, because it is easy to shake them off the fibers after drying. Carefully reheat the textiles and simmer for about half an hour; then take them out of the dye bath and you get the typical coffee brown. On silk the results are much lighter than on wool, going in the direction of beige. Add a shot of iron water to the dye bath for post-mordanting, and then put the selected dyeing fabric in the dye bath until you get the desired hue. This ranges between brown-green and dark green.

DYE SAMPLES

Wool

alum/
tartar

alum/
tartar
iron water

Silk

alum/
tartar

alum/
tartar
iron water

ROSELLE, KARKADE

(HIBISCUS SABDARIFFA)

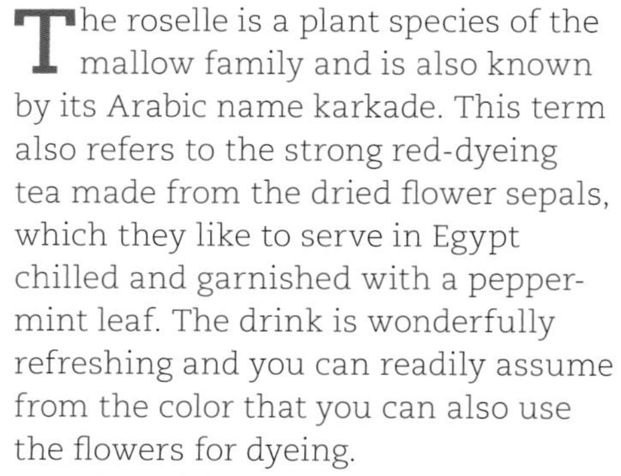

The roselle is a plant species of the mallow family and is also known by its Arabic name karkade. This term also refers to the strong red-dyeing tea made from the dried flower sepals, which they like to serve in Egypt chilled and garnished with a peppermint leaf. The drink is wonderfully refreshing and you can readily assume from the color that you can also use the flowers for dyeing.

The original home of the roselle plant is Southeast Asia, but the plant that is grown in Egypt and Sudan has gained a special importance due to the fact that the flowers are harvested to make tea. The roselle plant has also been cultivated in South America and is made into tea there. Hibiscus is an annual plant. Harvest the petals when the seeds are formed but have not yet ripened, and then dry them. The dried flowers can be enjoyed as a tea or used as a dyestuff.

Administered as a healing plant, roselle has an antibacterial effect and relieves cramps, and beyond this, the plant is also known as a thirst quencher. For dyeing, incidentally, you can also use the flowers of the

Hibiscus blossoms in the garden.

Wool and silk dyed with karkade.

common mallow (*Malva sylvestris*) to get the same dyeing results!

DYE RECIPE

For dyeing, both the commercially available dried flowers and fresh flowers can be used, but the same thing applies for these as for all blossoms from the dyer's garden: It is rare that you are able to harvest the quantity of fresh flowers you need for an entire dyeing all at once. Thus it is necessary either to dry them yourself or to resort to using purchased goods. The hibiscus flowers used for these dyeings actually come mostly from Egypt.

Use double the quantity of dried flower sepals by weight in proportion to dyestuff and soak these for a few hours, but you can also leave them in the water overnight. After soaking, boil the dye bath for half an hour, let it cool, and strain the flowers into a dye bag. Add this with the dyeing fabric to the dye bath and heat it slowly to simmering. Leave the dyeing fabric in the dye bath for a half hour to one hour. As with most plants that dye red, as an exception, I don't use a cold mordant, but rather pre-mordant with alum and tartar, since this makes the red dyestuff act with greater effect. The dyeing creates a soft pink on wool, and a rich dusky rose on silk. You cannot do a second dyeing with hibiscus flowers, since the color is almost completely exhausted after the first dyeing. Here you can really see the limited dyeing power of flowers, which have already exuded their brilliance into the world. It is however certainly possible to post-mordant; add a shot of iron water to the dye bath and immerse the dyeing fabric you want to post-mordant. On silk, you can expect intense dark gray shadings of color, which have a slight tinge of red. I'm not impressed by the results on wool, which tend towards gray-green.

Karkade tea with peppermint and dried flowers.

CARROT

(DAUCUS CAROTA SPP. SATIVUS)

DYE SAMPLES

Wool

cold mordant

cold mordant iron water

Silk

cold mordant

cold mordant iron water

The carrot has been known as a cultivated plant since antiquity. This tasty vegetable was already known in ancient Greece and ancient Rome, and by about 60 CE, the carrot was mentioned as a healing plant, with the author of the book emphasizing that wild carrots possessed much more healing powers than the cultivated ones. While in antiquity the wild carrot was used as an aphrodisiac, against stomach complaints or ulcers, today it is used to regulate blood sugar and also as diuretic tea.

Today we mostly know the orange carrot, but even as early as the Middle Ages different-colored varieties were known, from white to yellow and red to dark violet. Only recently, when we are again starting to value heirloom varieties and to plant them again, are we able to buy varieties other than the orange carrot. The wild carrot (*Daucus carota* spp. *carota*) is related to our carrot, because the carrot as we know it in our culture was bred from it. The roots of the wild carrot can be prepared like cultivated carrots and the greens look similar. The greens of both plants can be used for dyeing, and they yield the same results.

So anyone who likes to eat carrots and buys the carrot greens along with them from the farmer or at the health food store in the future doesn't have to throw away the greens anymore, but can use them immediately for dyeing. The carrot greens can also be dried, so you can collect enough during a whole summer season to work magic with the wonderful fresh yellow colors on silk and wool in the winter.

Wild carrots at the roadside.

DYE RECIPE

I myself dye exclusively with fresh carrots, because for me this is the quintessential dye for June, when the first fresh carrots come on the market. I pick up the greens in the market and at an organic farmer, where they are trimmed off when the vegetables are sold. For dyeing, I use double the quantity of carrot greens in proportion to the dyeing fabric, heat it in the dye bath, and let it boil for about two hours. Then I strain out the greens and add the pre-mordanted textiles to the dye bath, heat it carefully, and simmer for half an hour or an hour. The result is a bright spring yellow. It is possible to do a second dyeing, which yields a very light yellow. Post-mordanting the carrot-dyed fabric creates an appealing green on silk, which can best be compared to lime green; on wool the green color isn't as pronounced, but rather tends towards a yellow-green. Here, too, simply add a shot of iron water to the dye bath and re-immerse the previously removed dyeing fabric, which doesn't have to be heated any longer. For those who enjoy experimenting, it is also a good idea to post-mordant the second dyeing, which results in very light green tones.

Freshly dyed wool.

Wool and wild silk dyed with carrot greens.

TEA PLANT

(*CAMELLIA SINENSIS*)

DYE SAMPLES

Wool

1st dyeing
cold mordant

2nd dyeing
cold mordant

1st dyeing
cold mordant
iron water

Silk

1st dyeing
cold mordant

2nd dyeing
cold mordant

1st dyeing
cold mordant
iron water

The tea plant *Camellia* belongs to the camellia family and was originally distributed in Japan, China, and India, as well as in Laos, Myanmar, Thailand, and Vietnam. Today tea also flourishes in Africa, South America, Iran, and Nepal and is even scattered in some parts of Europe. Four varieties are known within the species *Camellia sinensis*, two of which are especially well-known and are widely distributed as cultivated plants, and hybrids of these two varieties are used for tea production.

Initially only green tea was available in Europe, while black tea reached Europe toward the end of the nineteenth century. The production of green and black tea differs significantly: For green tea, the leaves are harvested and immediately heat treated and dried, while for black tea, the slightly withered leaves are exposed to very moist air for a few hours, which causes slight fermentation. Then the leaves are likewise dried.

Tea drinkers certainly notice the fact that black tea, as well as coffee, dyes well on their teeth, which can become discolored by drinking a lot of it. Tea can also be used to dye textiles, for which we use black tea. It isn't necessary to use an expensive black tea to dye textiles; leftovers from the

Black tea.

Skeins of silk dyed with black tea.

kitchen work just as well for dyeing as freshly bought tea.

DYE RECIPE

To dye using black tea, use double the amount of tea by weight in proportion to the material. It's not important whether you mix different kinds of teas or tea leftovers, and it also is no problem if you add in the fruit- and herb-enhanced teas that have accumulated in the kitchen cupboard. Soak the tea for several hours and then boil it; let the dye bath cool, and put the tea into a dye bag. Add this to the dye bath together with the pre-mordanted dyeing fabric.

Heat the dyeing fabric slowly and simmer it for about an hour. On wool, the result is a delicate light brown and skin color from the second dyeing; on silk a delicate red-brown. If you post-mordant with iron water, you get a delicate green-brown, and on silk a dark brown with an intense green effect. To achieve this result, add a shot of iron water to the dye bath after removing the dyeing fabric and re-immerse the pieces that you want to post-mordant. The length of mordanting creates different shadings.

Although I mainly dye with wool and silk and very seldom use other materials, dyeing with tea is the big exception. A white cotton T-shirt dyed with black tea reaches a very nice range of skin colors, which I use for sewing dolls.

Wild silk scarves dyed with black tea.

DYE SAMPLES

Wool

1st dyeing
cold mordant

2nd dyeing
cold mordant

1st dyeing
cold mordant
iron water

Silk

1st dyeing
cold mordant

2nd dyeing
cold mordant

1st dyeing
cold mordant
iron water

ONION

(*ALLIUM CEPA*)

Onions find their place in every kitchen and are very easy to handle for dyeing. That is why for me, dyeing recipes using onion skins are among the classic beginners' recipes. The kitchen onion is one of the oldest cultivated plants on earth. It was used as a food by the ancient Egyptians and was also donated as a grave offering. The onion was a basic foodstuff for the Romans and they spread it throughout Europe. The health-promoting effect of the onion was also known very early; in the Middle Ages, onion amulets were even worn against the plague. Due to the proven antibacterial effect of the onion, it is even possible that these may have helped. In addition to its antibacterial effect, the onion also has a slight blood pressure-lowering effect and also reduces blood sugar slightly, is an anti-coagulant, and is anti-asthmatic. The onion is additionally used as a cancer prophylaxis or to support weight loss. In all these applications, the red onion is said to be more successful than the white or yellow varieties because it contains twice as many antioxidants. If you want to profit from all these benefits, it is important to only peel the onion

Yellow onions from grandma's garden, with plenty of onion skins.

Spun wool and silk cloth dyed with onion skins.

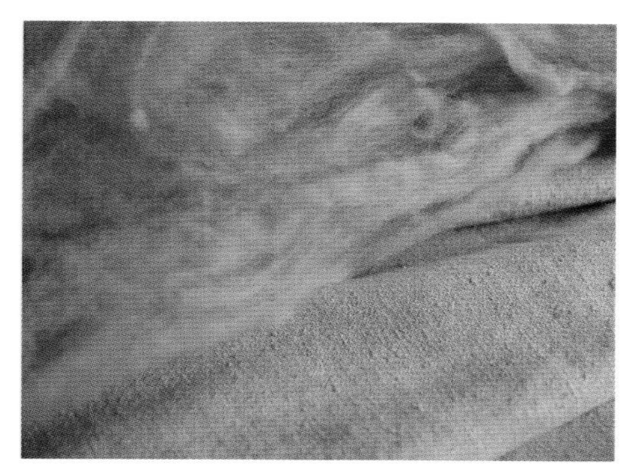

Unspun wool and wild silk dyed with onion skins.

a little, since the antioxidants are stored more intensively in the outer layers.

Onion skins give very beautiful warm yellow tones, and the second dyeing comes out somewhat lighter. You can also do a one-bath dyeing with onion skins. The material doesn't have to be pre-mordanted, but put the mordant directly into the dye bath. To do this, strain the onion skins into a dye bag and return this to the dye bath. Now add the calculated amount of alum to remove the onion skin and not the layers of the bulb. For dyeing I distinguish clearly between the red and yellow onion. Therefore, first do the recipe using yellow onion skins.

DYE RECIPE WITH YELLOW ONIONS

Onion skins are very light and it takes some time until you have collected enough for a dyeing. If you have a farmer nearby who still grows onions, you can ask him for the skins. You can also ask for skins at organic stores and markets. Onions from the supermarket are usually already peeled so well that there are hardly enough pickings for dyeing. The skins should be soaked overnight and boiled one to two hours the next day. Then let the onion skins cool, and either strain them into a dye bag, which is then returned to the dye bath, or use contact dyeing, that is, just leave the skins in the dye bath.

Put the onion skins, about the same amount by weight as dyeing fabric, in the dye bath, that is, the dyeing fabric is to be treated with 10% alum by weight. Thus, for 1 kg of dyeing fabric, dissolve 100 g of alum in warm water and add this to the dye bath. Now add the dyeing fabric, heat the dye bath again slowly, and simmer it for half an hour to an hour. Post-mordanting with iron water yields green hues in various shades.

DYE RECIPE WITH RED ONIONS

I am noting the red onion as a dyestuff separate from the yellow onion here, because in my opinion it hides some secrets. Like many plants that contain red dyestuff, onion skins dye green when this is done properly, but this dyeing often conceals itself and only comes to light when all the key data are right. And these are hard to figure out! When dyeing with red onion skins, as with other

Wool
cold mordant

Silk
cold mordant

Red onions and skins.

plants, things depend on the season and the origin of the dyeing fabric, but all this seems to be more important for onions. While one day I was dyeing, using painstakingly collected red skins from Italy because there were no native onions, and had to struggle to obtain this green, a few weeks later the domestic onions dyed this color flawlessly. On another occasion, the skins may have become too intense, since the result was a wonderfully rich dark brown, but no green wanted to develop out of it. Another time, the second dyeing yielded the hoped-for green tone. However, since all the shades of color from red onion skins are really beautiful, I would like to encourage you to experiment here, because every prospective dyer will always develop their own recipes.

My basic rule is to use the same amount by weight of red onion skins in proportion to the dyeing fabric. If you use more, the result will be darker and the green dyestuff may not emerge at all. So if you should determine that the skins you used yield very dark results, you can also try dyeing with half the quantity of skins by weight. Some dyer colleagues use only a 10% portion of skins by weight, as they already know their onions and know that they yield a very intense color. The green from the red onion dissolves out at a certain time in the dyeing process and the textiles that have taken on the green dyestuff should be removed from the dye bath immediately. The moment when the green turns into a dark brown is just a moment—if it has passed by, it's simply too late.

The mordant also has a big influence—I have found that materials that were only pre-mordanted for a short time can develop the green better than materials that were kept in the pre-mordant for a longer time. I never got this massive difference with any other dye plant!

You can therefore apply as **key data**:
• start with the same amount of skins, and
• only use fabric that has been pre-mordanted for one to two hours in the cold mordant.

Red onions dye silk green—a matter of the right timing.

Soak the onion skins overnight, and the next day add the fabric to the dye bath with the skins and heat it slowly. Remove the textiles at the time when they have taken on the desired hue; the dyeing goes from green to chestnut brown to dark brown. When doing this dyeing, you should stay right by your pot!

Unspun wool in various color shadings, dyed with red onion skins.

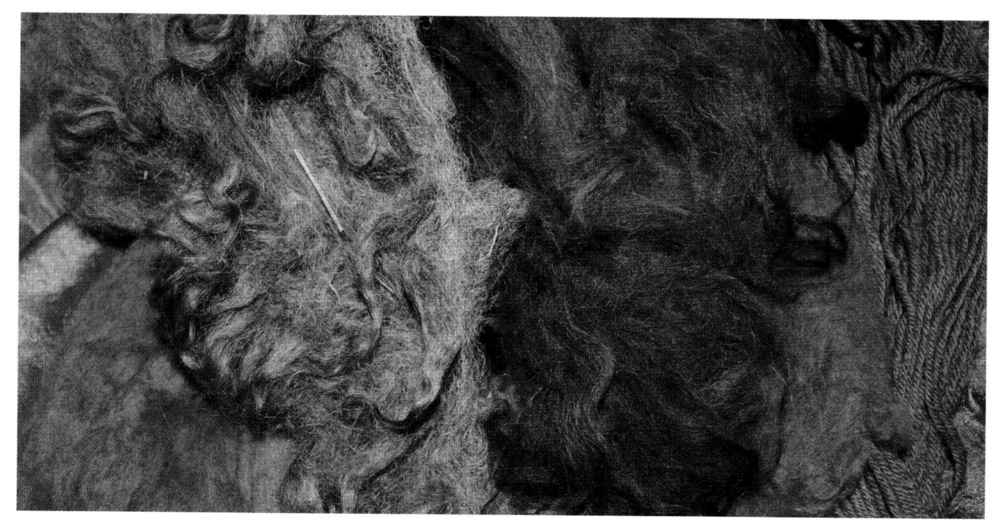

DYE SAMPLES

Wool

cold mordant
iron water

Silk

cold mordant
iron water

EXOTICS FOR DYEING

Dyeing with Special Plants and Insects

If you look back into the history of dyeing, or if you look at the plants described in this book, you might be tempted to describe many more dye plants as exotic, since indeed a considerable number of them originally came from another region of the world. Indigo comes from India and is still produced there, dahlias come from Mexico, hibiscus from Southeast Asia, and coffee and tea from Asia and Africa. I have gathered together under "exotics" all those dye plants and dye insects—for here we have included a non-plant dyestuff—that are especially exotic for me. Some are in this group because they are among the woods that we hardly use or no longer use anymore, such as red wood or bloodwood, or because they aren't of plant origin, or because we have integrated them so much into everyday use that we are hardly aware that they are exotics.

Brazilwood (*Caesalpinia echinata*), which is also commonly referred to as red wood, is one of the dyestuffs that are no longer used today, but were of great importance in the history of dyeing. Described correctly, Brazilwood belongs to a group of woods that, collected together, are used to dye under the term "red wood." Starting in the thirteenth century, red-dyeing woods were already being transported over the sea routes from India to Europe; *Caesalpinia echinata* has been used as a dyestuff since the fifteenth century. The tree got its popular name from its native country, Brazil, and in 1978 it was named the national tree of Brazil. Due to its dyeing properties, the stands of *Caesalpinia echinata* trees were reduced to such an extent that today it is under strict environmental protection and is therefore no longer available as a dyestuff.

In addition to the exotic plant dyes described here, we know of many others that still have their place for dyeing processes. These include, for example, the red sandalwood tree, the wood of which produces a delicate pink comparable to the results from

Cloth dyed with indigo.

Bloodwood.

karkade; the achiote, also called annatto; or the yellowwood, also called the dyer's mulberry tree. When dyeing with annatto, the seeds of the shrub are used and the results are intense orange, comparable with the results of dyeing using dahlias. The results with dahlias are moreover more lightfast than those from annatto. For yellowwood, as with Brazilwood or red wood, the wood shavings are used for dyeing and the results are different gradations of yellow hues. Since many of our domestic plants dye yellow, you can confidently do without this exotic wood.

Dried cochineal insects are available from specialist retailers.

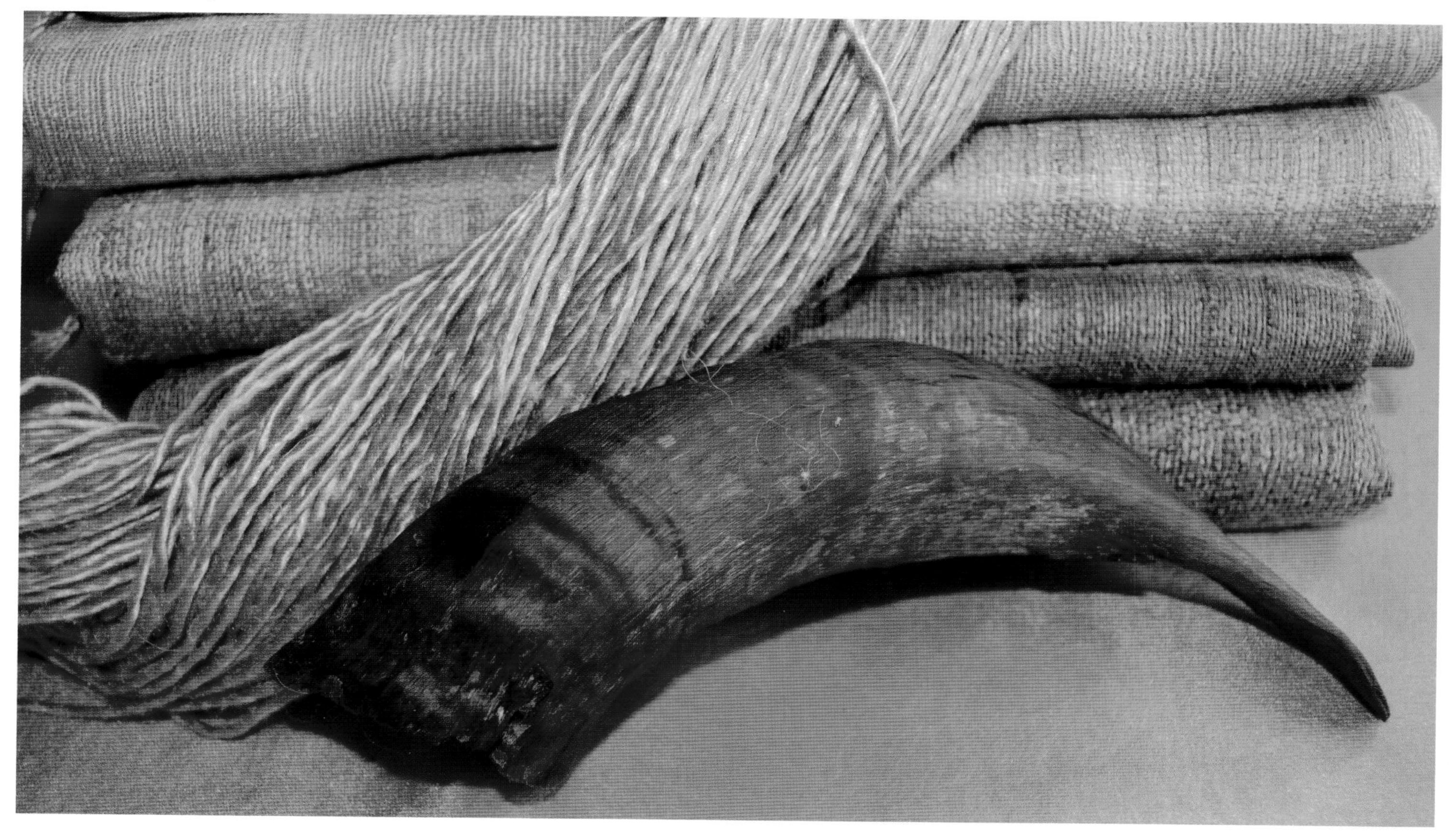

Textiles and horn dyed with cochineal.

BLOODWOOD, CAMPECHE TREE

(HAEMATOXYLUM CAMPECHIANUM)

Wool

alum
tartar

cold mordant

Silk

alum
tartar

cold mordant

Bloodwood is the heartwood of the campeche tree, which is native to Central America. It got its name of campeche tree from its original home, the state of Campeche in Mexico. The bloodwood tree grows 10 to 33 feet high and can appear both as a shrub and a tree. The dyestuff hematoxylin, which is contained in the wood, forms the actual dyestuff hematite when it is exposed to air. We can reinforce this process by fermentation in the air. As a dyestuff, you can get the already crushed, treated, and dried wood in a specialty store.

Dyeings done with bloodwood are a bit outside the range of the palette of the colors I dye with. Even though you can obtain the entire range of purple hues, even up to a very dark blue with bloodwood, I only use this wood to overdye, especially if I want to get black.

DYE RECIPE

Apart from the fact that dyeing purely with bloodwood doesn't fit within my preferred color palette, I don't particularly like the wood because it makes the fibers very brittle. Bloodwood is, however, an important partner for overdyeing, because it opens up the dark tones of the color palette. To dye with bloodwood, soak the dyeing fabric for at least two to three days, using half as much dyestuff as dyeing fabric. Thus, for 100 g of wool, use 50 g of bloodwood; if you want to get lighter or darker results, vary the proportions accordingly. Boil the dye bath for three to four hours, then let it cool, and strain the dyestuff into a dye bag. Put this back in the dye bath with the pre-mordanted dyeing fabric, and simmer this for about half an hour, until you get the desired hue. Here, of course, it is possible to do a second or third dyeing with correspondingly lighter color results. This dye bath works just as well for overdyeing cochineal (result is a dark

Bloodwood-dyed mordanted wild silk.

Black silk scarf, pre-dyed with indigo, overdyed with bloodwood.

purple) or various hues of yellow (the results are the most diverse green tones).

For overdyeing fabrics dyed with nut or indigo, use twice as much bloodwood by weight in proportion to the dyeing fabric. Prepare the dye bath in the same way as described, but add some yellow-colored flowers like camomile or marigold to it. The yellow dyestuff enhances the development of the black hue—this insider tip from the dyer witches' kitchen produces results that can really be seen. Immerse the dyeing fabric, simmer for an hour, and then leave it for a few hours in the dye bath. The results are nut black or indigo black, which means that the black shades shimmer lightly in the basic color. If you subsequently add iron water to the bloodwood bath and re-heat the already dyed materials in the dye bath, you can obtain a rich black. For anyone who likes to experiment, it is a good idea to dye some undyed but pre-mordanted material in the dye bath before adding the iron water. The results come out very differently. I have already dyed beautiful blue tones on wool or horns with this method; these I simply put in the dye bath without pre-treatment.

Wool and horn dyed in a black dye bath.

OVERDYEING SAMPLES

COCHINEAL INSECT

(*DACTYLOPIUS COCCUS*)

Wool

1st dyeing
10% alum
10% tartar

1st dyeing
10% alum
5% tartar

2nd dyeing
10% alum
10% tartar

Wool

1st dyeing
10% alum
10% tartar
iron water

1st dyeing
10% alum
5% tartar
iron water

2nd dyeing
10% alum
10% tartar

Cochineal is the only non-plant dye agent described in this book. The cochineal insect originally comes from South and Central America and lives there on opuntia, a plant genus of the cactus family. The dyestuff carmine can be extracted from the female insects.

The cochineal insect has long been used to produce red dyestuff and the oldest finds date from pre-Christian times and were discovered in Peru. It has not been proven how long these scale insects had already been cultivated, but by the Colombian period, opuntia, on which the insects are raised, was already being cultivated. The cultivated specimens are up to twice as large as the wild ones and provide correspondingly more dyestuff. The red dyestuff was already extremely valuable by Aztec times, just as valuable as the textiles dyed with it. Records from the sixteenth century indicate how much cochineal the peasants had to hand over to the Aztec rulers. In Europe, dyeing with Polish cochineal was known at this time, but the knowledge of such dyeing was reserved to a few guilds. In the sixteenth century, cochineal was known in Europe and it was also recognized that dyeing with these insects was much simpler than dyeing with Polish cochineal. Today, you can get the dried insects from retailers of dyeing supplies.

Wild silk scarves dyed with cochineal.

DYE RECIPE

When dyeing with cochineal, the materials will take on the color more intensely if they are pre-mordanted with alum and tartar, but on silk I have already achieved beautiful results with cold mordants. Pre-mordanting with alum and tartar is described in detail in the section on mordanting. To dye, crush the cochineal in a mortar to get 25% in proportion by weight to the dyeing fabric, or grind it finely in a coffee grinder, then stir it into a small pot together with 10% tartar and 250 ml of water, and let stand for a few hours. Thus, for 1 kg of wool you would need 250 g of cochineal and 100 g of tartar. Then simmer the mixture for half an hour. Next strain out the scale insects into a stocking, tie it closed, and then add it to a larger dyeing pot with a corresponding amount of water (enough so that the dyeing fabric is well covered). Add the pre-mordanted material to this dye bath, heat it slowly, and keep it at a boil for a half hour to an hour. Cochineal has a great many varieties to offer and yields copious results; at the concentration stated, it is possible to make many secondary dyeings, especially on silk, which come out correspondingly lighter. You can also experiment with the dosage of tartar, or leave out the tartar altogether, and then you get a slightly blue-tinged pink.

By overdyeing textiles dyed with madder, you obtain a rich raspberry red, a hue that it is barely possible to create with natural agents. It is also possible to post-mordant with iron water, which produces intense purple tones, as does overdyeing cochineal-dyed fabrics with indigo. For post-mordanting, add a shot of iron to the cooled dye bath, then re-immerse the desired textiles in the dye bath, and take them out when they display the desired color result.

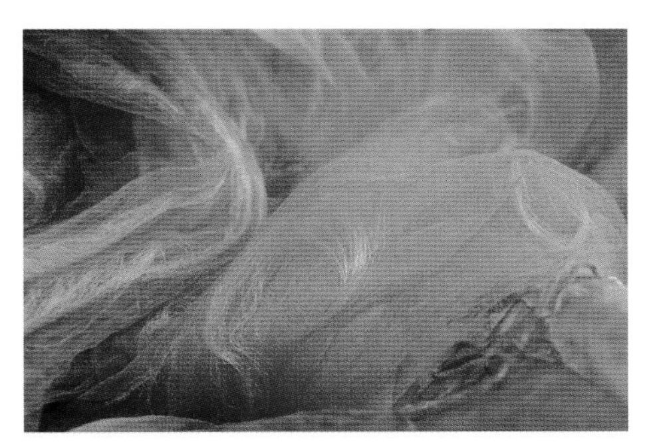

Silk dyed with cochineal.

Wool, spun and unspun, and silk, dyed with cochineal.

DYE SAMPLES

Silk

1st dyeing
10% alum
10% tartar

1st dyeing
10% alum
5% tartar

2nd dyeing
10% alum
10% tartar

Silk

1st dyeing
10% alum
10% tartar
iron water

1st dyeing
10% alum
5% tartar
iron water

2nd dyeing
10% alum
10% tartar

DYEING WITH COCHINEAL

The dyestuff: dried cochineal scale insects.

The dried cochineal scale insects are ground in a coffee grinder or finely crushed in a mortar.

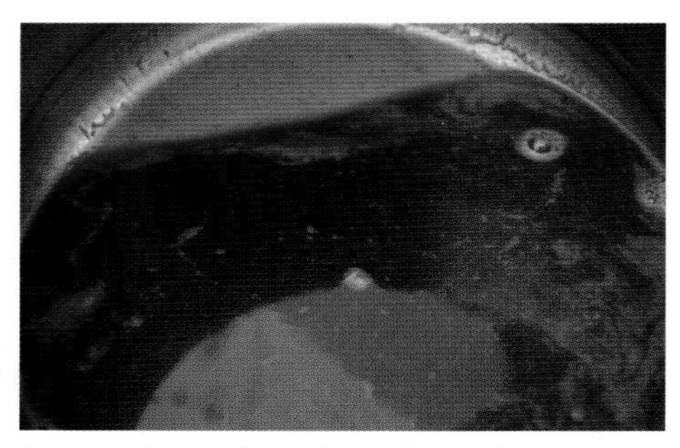
Simmer the cochineal powder with tartar in a little water.

Strain the dissolved and boiled dyestuff into a fine stocking.

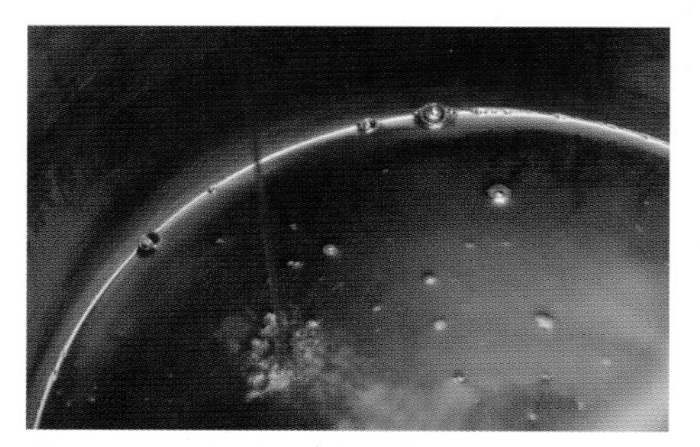
The orange dye bath dyes from magenta to pink.

The dyed fabrics dry in the sun.

Left: Wild silk scarves after being dyed with cochineal.
Right: The textiles dyed as described in the recipe—you can see the results!

HENNA

(*LAWSONIA INERMIS*)

DYE SAMPLES

Wool

without mordant

cold mordant

without mordant iron water

cold mordant iron water

Silk

without mordant

cold mordant

without mordant iron water

cold mordant iron water

The henna tree is a plant with many special qualities. On the one hand it is the only plant of the genus *Lawsonia*; no other plants on Earth have been assigned to this genus. On the other hand, to date the original source of the plant has not yet been established, since it has been cultivated in many areas for thousands of years. The henna tree was already known in ancient Egypt and is also mentioned by the Greeks of antiquity, mainly because of the beguiling fragrance of its flowers. This scent is one reason why the tree was cultivated in many areas very early on. Another reason is the henna powder extracted from the dried leaves. It was already common to dye hair and paint skin with henna in ancient Egypt and this has been detected on mummies; here, applying henna to the skin took on a much more important role. Painting the skin served mainly ritual purposes; it was considered a way to deflect the evil eye. Painting the hands and feet artistically with henna is still common in the countries of North Africa and the Middle East as well as in India, while in the West henna is mainly used for dyeing hair. Depending on the quality, the powder produced from the leaves of the henna tree can create different red tones ranging to brown; all commercially available "henna powders" that promise other color

Henna powder.

results are mixtures with other plants or mixtures with chemical additives. Traditionally, indigo is added to henna powder in North Africa and the Middle East to give a black dye. In addition to skin and hair, henna can also be used to dye textiles that contain protein, and the result is a bright, luminous yellow-brown to red-brown to brown on wool, as well as surprising yellow tones on silk.

DYE RECIPE

For dyeing, mix the same amount of henna by weight as the dyeing fabric with a little water so that no lumps form, then put this paste in the dyeing pot and stir the mixture well. Next simmer the dye bath for an hour before you strain out the powder through a coffee filter. Any other way of straining the powder isn't fine enough, and the henna powder will just pour out with the dye bath. You can also leave the powder in the dye bath, because you can readily shake it out after the dyeing fabric is dried. Leave the dyeing fabric to dye in the dye bath for a half hour to one hour. Even without pre-mordanting, henna produces beautiful results on wool; but as is the case for very few dyestuffs, without a pre-mordant, it has no effect on silk. To post-mordant with iron water, add a shot of this post-mordant to the dye bath and re-immerse the already dyed skeins of wool and silk in the dye bath. The results on wool extend to brown and ocher, and a dark olive green is created on silk.

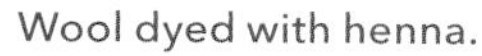
Wool dyed with henna.

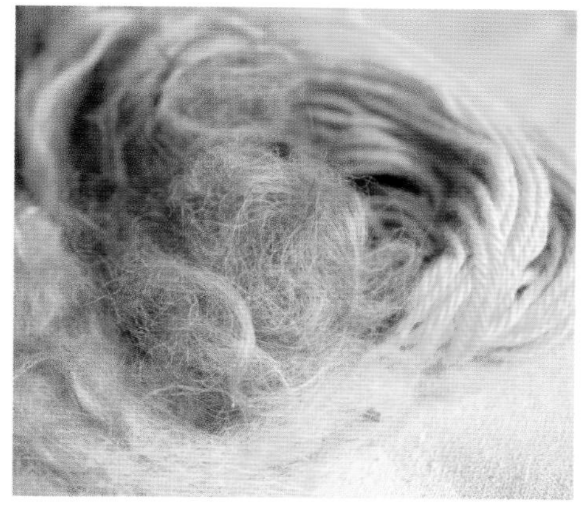
Wool dyed with henna.

Skeins of wool dyed with henna.

INDIGO

(INDIGOFERA TINCTORIA)

DYE SAMPLES

Wool

1st dyeing

2nd dyeing

3rd dyeing

Silk

1st dyeing

2nd dyeing

3rd dyeing

Indigo—the great mystery of the dyer's world, the substance that led to the decline of an entire economic branch in Europe and turned the dyer's world of Central Europe upside down. Dyeing with indigo is an elaborate process and is embedded differently in the tradition of every country whose people know indigo. In Europe, up to the seventeenth century blue dyeing was done using dyer's woad. Although indigo had been known in Europe since ancient times, little was available and large amounts arrived in Europe only after the sea route was opened up to India. Compared to dyer's woad, the indigo plant produces thirty times more dyestuff, so producing blue dye from woad became unprofitable in Europe by the seventeenth century.

Within a short time, a strictly regulated branch of industry went under, and whole villages that had been committed to cultivating woad were impacted. It wasn't until 1897 that blue dye was produced synthetically in Europe and indigo was gradually forced off the market.

The bright pink-flowering indigo plant is grown on a limited scale in the United States, and thrives wonderfully well in Europe and similar latitudes, but the dyestuff it contains is by no means as intense as that from the indigo plants that thrive in India. To extract the dye from the fresh plant requires a huge amount of plants, and since it is a very laborious process, it is advisable to

Genuine indigo also grows in home gardens.

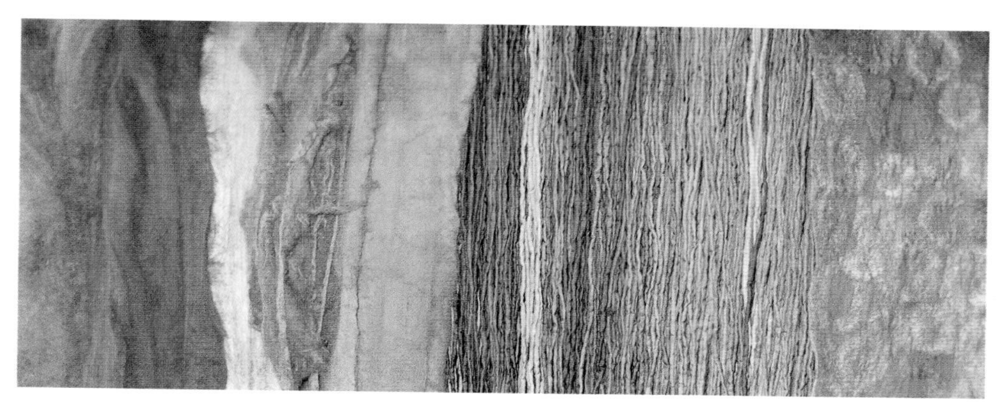

Various materials dyed and overdyed with indigo.

buy your indigo dyestuff in powder form. You can't use the basic recipe for dyeing with indigo because it requires a special chemical process to get the blue color into the material. We are talking here about so-called vat dyeing.

DYE RECIPE

For dyeing, use 20 g of indigo powder, 40 g of hydrosulfite (twice), and 40 g of soda for about 2 kg of material. These quantities can, of course, be altered as desired, as long as you keep to the quantity ratios. The given quantity yields optimal color results for 2 kg of material. If you only want to dye three small silk scarves, use correspondingly smaller vats.

A thermometer is indispensable for the indigo vat, since the vat must never get too warm. First, dissolve the soda by stirring it into 2 liters (0.53 gallons) of water that has previously been heated to 104°F (40°C). Then add a portion of hydrosulfite, and finally add the indigo powder carefully. This "stock vat" is slowly heated to 131°F (55°C). In a large pot—for this, containers that are used to keep tea or mulled wine warm work particularly well, since they can be kept at the desired temperature—heat another 20 to 25 liters (5.28 to 6.60 gallons) of water to 104°F (40°C). Then add the second dose of hydrosulfite to this water. Next, add the contents of the stock vat in such a way that it absorbs as little oxygen as possible. Hold the pot very low over the water when you pour out the liquids together. Then re-heat this vat to 131°F (55°C). As soon as a bluish-green film forms on the surface, the so-called indigo flower, you can start dyeing. Dip the un-pretreated textiles in the vat, where they should always be completely covered by the liquid; when you take them out, wring out the textiles already right over

DYE SAMPLES

Overdyed cochineal–indigo

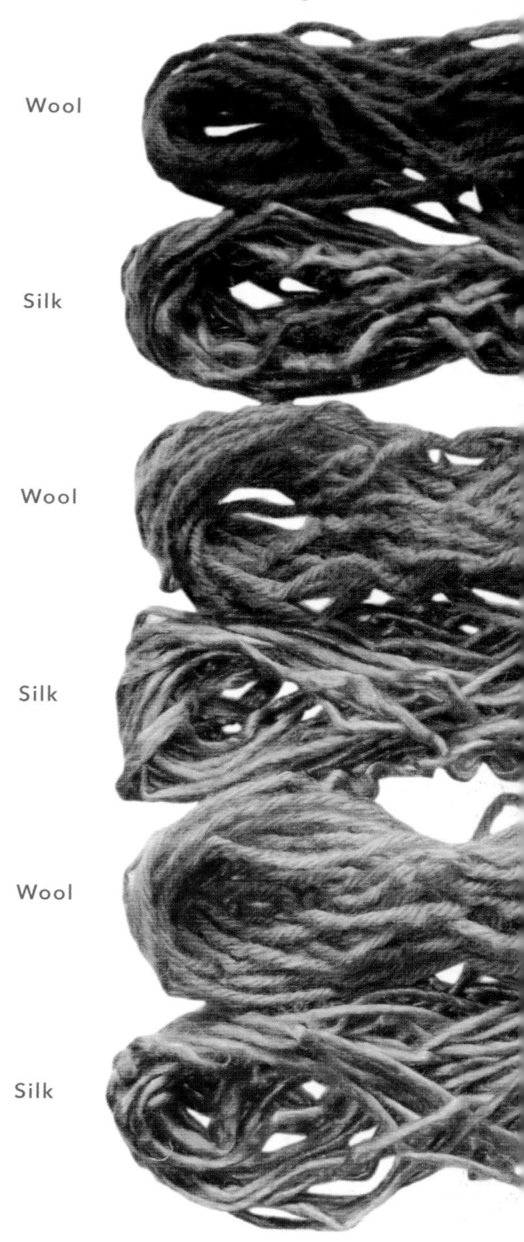

Wool

Silk

Wool

Silk

Wool

Silk

DYE SAMPLES

Various overdyeings

Marigold - indigo
Wool
Silk

Wild indigo - indigo
Wool
Silk

Birch - indigo
Wool
Silk

Horsetail - indigo
Wool
Silk

Wild silk scarves overdyed blue or green with indigo.

Wool dyed and overdyed with indigo.

the surface so that as little oxygen as possible gets into the vat.

The first dyeings are the darkest, and then you can observe how the color results become ever lighter. While the first pieces stay in the vat for only a few minutes and show a dark blue color, for later dyeings, you definitely can leave the textiles in the vat for two hours. If you aren't happy with a result, put the piece back in the dye bath again and after-dye it. If you only get very light pale blue, the vat is depleted. But you can still "re-sharpen" it up to the end by adding still more hydrosulfite. Thus, if the results come out too light, add a little hydrosulfite to the vat and this may again improve the color results.

You can create a large portion of the color palette by overdyeing with indigo. By dyeing fabrics that have already been dyed pink with cochineal, you can obtain the most diverse purple hues, and all textiles previously dyed yellow can take on different green variations from overdyeing with indigo. The examples illustrated in the dye samples show the gradations that you can achieve as the dye bath constantly gets weaker. You can also produce gradations depending upon how long you leave the textiles in the dye bath, and it is also possible to do a second overdyeing to improve the results.

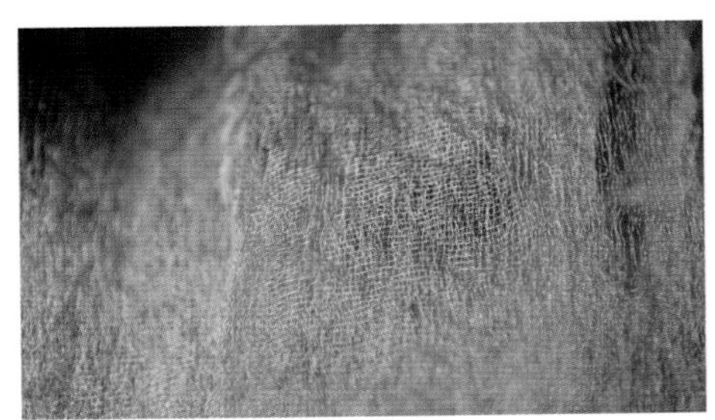
Wool scarf (top) and wild silk (bottom), dyed with indigo.

DYEING WITH INDIGO

Materials for dyeing with indigo.

Stir in the indigo powder.

Pour out the combined liquids carefully.

Use a cooking spoon handle to stir the skeins of wool (covered, if possible) in the vat.

After you remove it, thoroughly wring out the material right away.

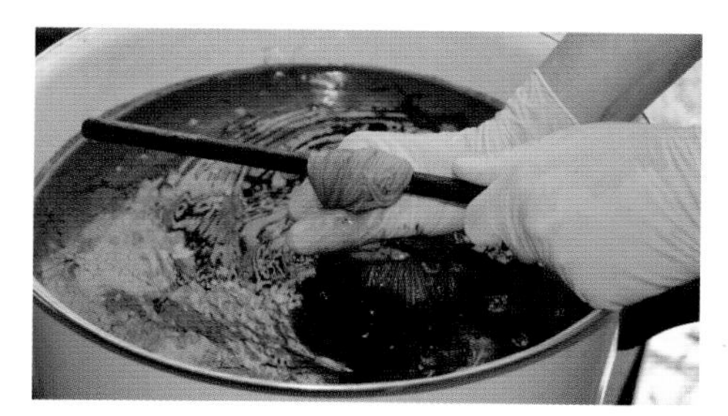

The blue color develops through contact with oxygen.

Putting yellow pre-dyed wool in the indigo vat

Overdyed yellow: by waving the fabric in the air the green color immediately comes into its own.

Silk pre-dyed pink with cochineal . . .

. . . goes into the indigo vat and comes out violet.

CATECHU, CUTCH TREE

(*ACACIA CATECHU*)

DYE SAMPLES

Wool

without mordant

without mordant potash

without mordant iron water

cold mordant

cold mordant iron water

Silk

without mordant

without mordant potash

without mordant iron water

cold mordant

cold mordant iron water

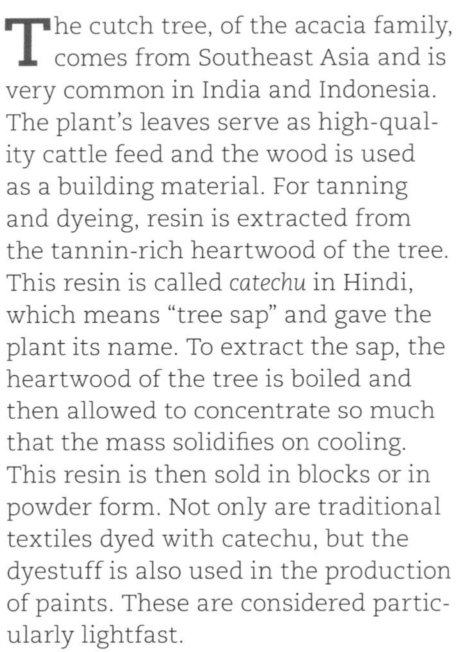

The cutch tree, of the acacia family, comes from Southeast Asia and is very common in India and Indonesia. The plant's leaves serve as high-quality cattle feed and the wood is used as a building material. For tanning and dyeing, resin is extracted from the tannin-rich heartwood of the tree. This resin is called *catechu* in Hindi, which means "tree sap" and gave the plant its name. To extract the sap, the heartwood of the tree is boiled and then allowed to concentrate so much that the mass solidifies on cooling. This resin is then sold in blocks or in powder form. Not only are traditional textiles dyed with catechu, but the dyestuff is also used in the production of paints. These are considered particularly lightfast.

The cutch tree also has traditional importance as a medicinal plant, and the tannin-rich wood has an astringent effect; it has been used in Indian medicine for years as a remedy for inflammations, especially of the mouth and throat. Catechu is also administered for diarrhea. In Ayurvedic medicine, not only the plant resin is used but also the bark and flowers.

DYE RECIPE

Add half the amount of catechu powder in proportion to the dyeing fabric to 1-2 liters (0.26-0.53 gallons) of water, stir well, and heat carefully. Here, don't use the entire amount of water that you would normally use for dyeing because it is

Catechu powder.

important to make sure that no sediment is formed during the heating. While you are warming the liquid, it is especially important to keep stirring so that the catechu powder doesn't stick or burn. Let this mixture simmer for half an hour while watching it, and only then add the rest of the water you need to cover the dyeing fabric completely with the dye bath. In contrast to henna powder, catechu dissolves completely and therefore doesn't have to be strained out. Add the dyeing fabric to the cooled dye bath and heat it again slowly; let it simmer for half an hour to an hour and you will obtain red-brown to orange-brown. You can even dye un-mordanted materials with catechu!

Catechu produces wonderful nuances as a result of post-mordanting with potash. Post-mordanting with potash makes sense above all if you want to intensify yellow tones in a dyeing, so I use this method only for selected dyeings. To post-mordant with potash, take the dyeing fabric out of the dye bath and set aside briefly. Now dissolve one tablespoon of potash per 500 g of dyeing fabric in some water and pour it into the dye bath. Add the dyeing fabric to this post-mordant and orange-brown hues develop! The color nuances can vary depending on the fabric and how long it is left in the post-mordant. Post-mordanting with iron water also produces lovely results; on wool, you get shadings of brown and on silk dark brown to green-brown.

Wool and silk dyed with catechu.

Skeins of silk dyed with catechu.

DYE SAMPLES

TURMERIC

(CURCUMA LONGA)

Wool

cold mordant

cold mordant
iron water

Silk

cold mordant

cold mordant
iron water

Turmeric, also called saffron or yellow ginger, belongs to the family of ginger plants. You can recognize its relationship to ginger by the very special sharpness of turmeric powder. The powder is produced from the rhizome, the plant's shoot source that develops under the soil. It looks very similar to that of ginger, but is dark yellow, and not light colored as ginger is. The egg yolk yellow turmeric powder has been used as a medicine and a remedy in India for more than 4,000 years and is highly valued in Ayurvedic medicine. Turmeric has a strong antioxidant effect and is administered against joint inflammations, and the plant is also supposed to be effective against arthritis and to reduce the risk of stroke and heart attack. This healing plant with all its many modes of action would be worth a book on its own, because it has so many potential uses!

Used as a toothpaste, the egg yolk yellow powder ensures that tooth

Turmeric in powder form and in pieces.

Silk dyed with turmeric.

Unspun wool dyed with turmeric.

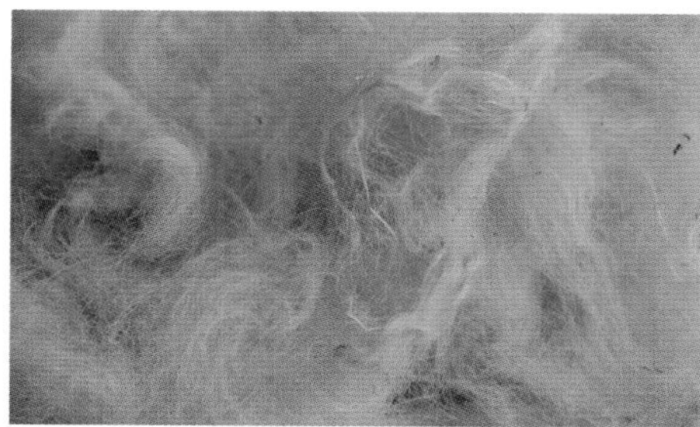

discoloration caused by smoking, coffee, or excessive tea drinking is reduced within a few weeks. To do this, just dip your moistened toothbrush into the powder and brush your teeth as if you were using toothpaste. While it whitens teeth, turmeric dyes everything else yellow, and this is where it finds its use as a food coloring in curry mixtures or in mustard and pasta. And of course you can also dye textiles with turmeric!

DYE RECIPE

For dyeing, use the same amount of turmeric powder in proportion to the dyeing fabric and mix with a little water into a paste. Thus, use 100 g of turmeric powder for 100 g of wool.

Add another liter (0.26 gallons) of water to the powder and water mixture, stir the paste well, and simmer the liquid gently for an hour. The resulting fragrance delights the dyer's heart as much as the cook's! Then allow the dye bath to cool down some and strain out the powder, preferably through a nylon stocking pulled over a sieve. You should filter out the turmeric dye bath anyway, since the powder sticks very stubbornly to the fibers and can't be easily shaken out like henna. Now put the pre-mordanted materials in the strained dye bath, heat carefully, and simmer for about an hour. Turmeric produces wonderful intense yellow hues. Turmeric dyeings are not particularly lightfast and the textiles already get bleached out after a few months. Thus, dyeing with turmeric doesn't work well on fabrics you use every day.

OTHER DYEING TECHNIQUES

Eco Printing and Solar Dyeing

Textiles can be dyed in many different ways. The dye recipes described work particularly well for wool and silk, but you can definitely venture on using cotton, linen, or other materials. It is also very popular today to dye using plants as a part of "upcycling," which means enhancing outdated or faded clothing by refining it artistically, as we know from using the batik technique in the 1970s. So-called eco printing also comes out of the upcycling scenes, and it is gaining more and more friends, especially in the English-speaking region.

ECO PRINTING

Eco printing is nothing other than the printing on textiles using pieces of plants. Thus, plants aren't boiled in a dye bath, but caused to leave their impression on the fabric by using a particular binding technique.

Eco printing isn't difficult and always creates surprising results. You can use this technique to decorate cloths and scarves, but also on clothes that you want to enhance. Motivated dyers tell us in their various blogs how they dye almost everything from socks to a summer dress. Before you start on large pieces, you should make your first attempt on a sample piece to learn the technique.

For eco printing you need:

- Textiles you want to dye
- Various rollers of plastic, wood, or metal on which you can wind up the fabrics
- A bucket to soak the plants
- Plastic film for covering and rolling up, for example as it is used to cover up paintings
- Cord for binding
- Plants for printing

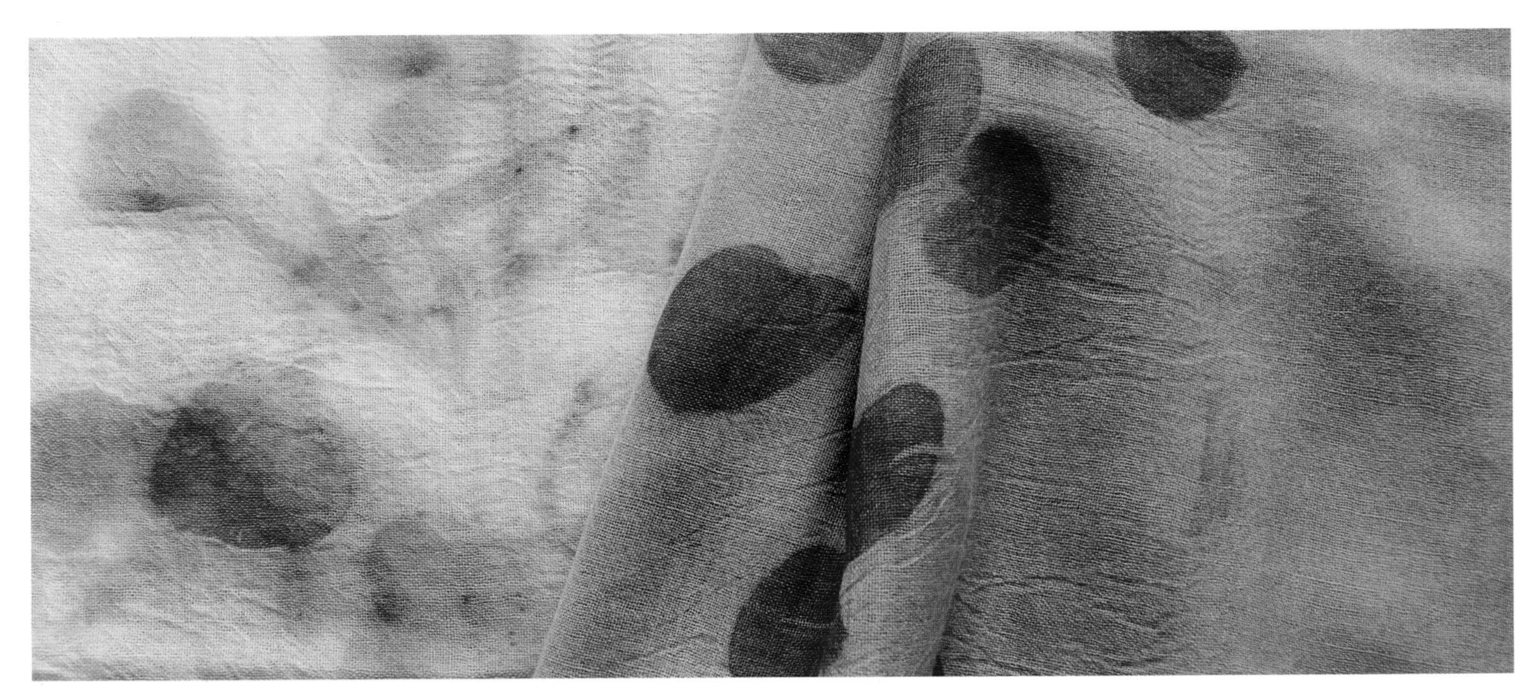

Scarves dyed with eco printing technique.

Various rollers for rolling up the textiles.

Eco printing works well both with classic dye plants and also with leaves from trees and shrubs, as we find them in parks and in the forest, just like grasses and flowers. Exotic plants like red maple or wisteria especially yield impressive results. Both fresh and dried plants can be used.

Step by Step

Start mordanting in a bucket. Here you can either use iron sulfate, which can be purchased from a dyeing accessory retailer, but in our sample we have simply used iron water because you can make this yourself, as described in the section on mordanting. Fill a bucket with 5 liters (1.32 gallons) of water and add a teaspoon of iron sulphate or a good shot of iron water. Eco printing doesn't follow any clear rules; you can try different variants. It's possible, for example, to add oak gall powder to the mordant, which means that the selected fabrics don't have to be pre-mordanted and the colors become darker and more intense.

Note here as a **reference point**: If the selected textiles are not pre-mordanted, you should add oak gall powder to the mordant for the plant pieces, at the proportion of about one tablespoon to 5 liters (1.32 gallons) of water.

The selected leaves are soaked in a mordant.

Lay the pre-mordanted leaves on the silk cloth.

Now add the selected plant pieces to the bucket; these must be well wetted with mordant, but it isn't necessary to soak the plants for an extended time.

Before you get to the creative part, you have to dampen the selected materials. The textiles should be well dampened but not wet at all, as when you take them well-wrung out of the washing machine or hand-washing basin. If the fabrics are too wet, the impression of the plant becomes less intense. Now lay the selected piece on plastic foil. The leaves, grasses, and flowers, well dampened with mordant, are placed on the fabric, and then there are no limits to your creative designs!

Next lay a piece of plastic foil, which should be at least the same size as the selected piece, over the finished leaf pattern, and then more leaves and flowers are laid on top of this. Now choose a roller, place it at the beginning of the fabric, and roll up the fabric and leaves tightly on this roller. Don't roll up the bottom plastic layer with it! Bind up the roller with fabric and leaves well with a piece of string, as shown in the picture below. If you use a very wide piece of fabric, you should first cover only half of it with leaves and fold the other half over the designed side. When rolling it up, you have to keep to the width of the roller you have available.

There are several potential ways to further treat the material. You can put the rolls with the rolled up fabric in a pot of water, heat it carefully as when dyeing, and simmer the immersed rolls for at least two hours. The longer the material is simmered, the more intense the impression of the plants.

A second possibility is treating the materials with steam. An electric pot works well for this, like those used to keep drinks warm at the Christmas market. Place the rolls on the rack in the pot and likewise expose them to steam for at least two hours. You can also steam the material in a conventional cooking pot. For this, put a rack, such as from a grill, on top of a can in the pot. Fill the pot with water

The material is laid on top and rolled up.

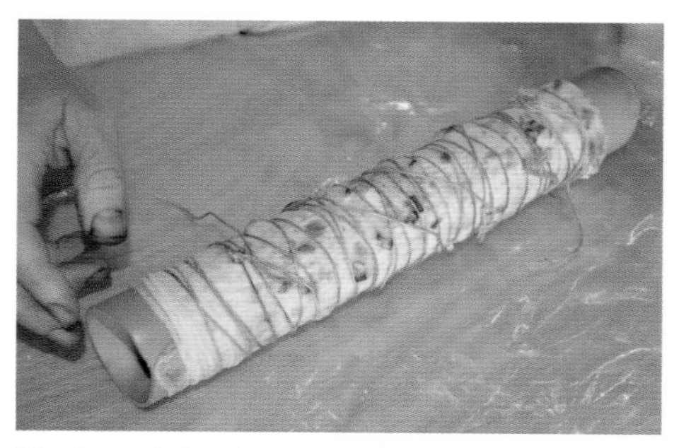
Tie the roll firmly.

The rolls are treated with steam.

Eco printing—rolls in a cochineal dye bath.

just so that the rack remains above it, lay the rolls on top, bring the water to a boil, and leave the fabric rolls to steam for at least two hours.

Of course, the rolls can also be additionally dyed. To do this, you simply put them in your dye bath of choice and dye as usual. After a boiling time of at least two hours, take the rolls out of the steamer or pot, cut off the string, and unroll the fabric from the roll. Now the wonderful plant impressions emerge—an experience like Christmas! Eco printing is always a surprise; you can never tell beforehand just how the result will turn out.

After the steaming time, the roll is unpacked.

Drying on the laundry line.

SOLAR DYEING

Solar dyeing is an incredible amount of fun, especially for children. Watching how the sun and warmth dye wool is a unique experience. Solar dyeing is a very special form of contact dyeing.

You don't need much equipment for dyeing with pure solar energy:

- Disposable glass jars with lids; the size depends upon the amount of dyeing fabric
- Un-mordanted wool fleece or skeins of unspun wool
- Plants from your dyer's garden
- Alum powder for mordanting

Plants that work especially well for solar dyeing include hollyhocks, dyer's chamomile, Mexican marigold, St. John's wort, common marigold, all kinds of onion skins, as well as various leaves and flowers you can use for experimenting. Of course you can also use dried dyestuffs such as madder root and bloodwood.

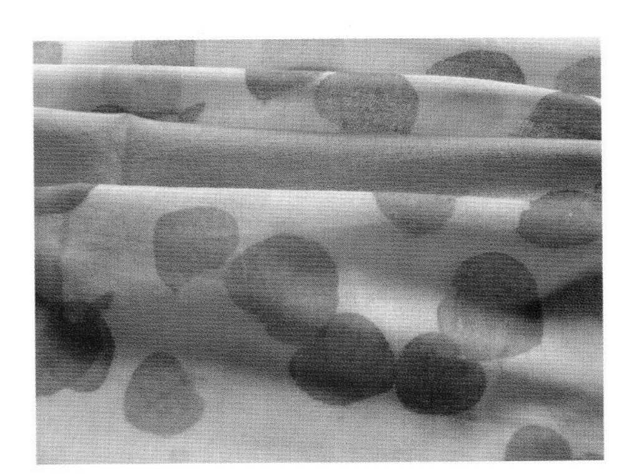

Eco printing with eucalyptus leaves.

Step by Step

Start with the selected plant pieces, now layer them alternately with wool and dyestuff in the disposable glass jar. Here you can choose freely according to your desire and mood. If you want to achieve a certain color, choose only one plant that dyes this color per jar. Thus, common marigold and St. John's wort work well together to achieve a magnificent yellow; madder root and dyer's chamomile for an intense orange and hollyhock for green. Now dissolve a teaspoon of alum per liter (0.26 gallons) of water in the liquid and empty it into the full-layered glass jars. Seal the jars and put them in a sunny place. There, you can watch as, from hour to hour, more of the color is transferred from the dyestuff to the wool. You can leave the jars in the sun for between one and four weeks. Every now and then, you should check

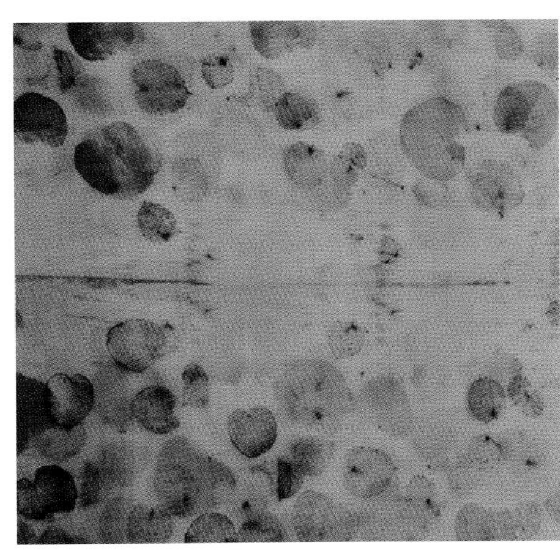

Eco printing with various leaves.

Hollyhock and maple are layered with wool.

The finished filled jars in the sun.

Alum water is poured over the finished layering.

After one week, you can already see the color well.

The pieces of plant are removed from the finished dyed wool.

The wool is hung up to dry.

The skeins of wool are ready to spin.

whether the dyestuffs and the wool are still covered by water, since they can start to mildew.

After the dyeing time, take the material out of the jar and remove the plant pieces as far as possible. Small pieces and fibers can be shaken out easily after drying. Let the wool dry out well in the fresh air and then you can treat it further, for producing wool pictures, for felting, or for spinning.

You can be playful with solar dyeing—especially when you dye with children. You can also add strongly colored flowers to the dyeing, even if they don't have any big effect. Here the visual experience when filling the jars comes to the fore. As long as there is also a real dye plant in the jar, you will also get the dyeing result generated by this plant. If you are pleased with solar dyeing and do a lot of dyeings, you can also put your chosen dyestuffs in tea strainers. Then you don't have to get the small plant pieces out of the wool, and the dyeing result is the same.

All the splendor of the garden goes into jars.

The colorful dye mixture from the garden.

Wool with dyestuff in tea strainers, a colorful mixture for the eyes.

DESIGN WITH DYED WOOL

Creative Gifts for Special People

WOOL PICTURES—PAINTING WITH WOOL

Wool pictures are a wonderful way to work the self-dyed woolen fleece into lovingly designed works of art. To create wool pictures, you should choose simple designs at the beginning, so that you can have a successful experience. More complicated designs require some practice. Beginners are well served if they use templates such as a drawing or a photo.

To make wool pictures you need:

- Unspun wool in various colors and qualities
- Several felting needles
- A piece of felt the size of the desired picture
- A foam pad

Step by Step

First, pin the cut piece of felt to the backing with one or more felting needles. If you want to create your wool picture, as in painting, with depth and a lot of structure, it should first be "primed" with white wool. But beginners can leave this step out. Pluck the white wool from the wool fleece in very fine pieces and lay it on the felt backing, and in this process you can already define a

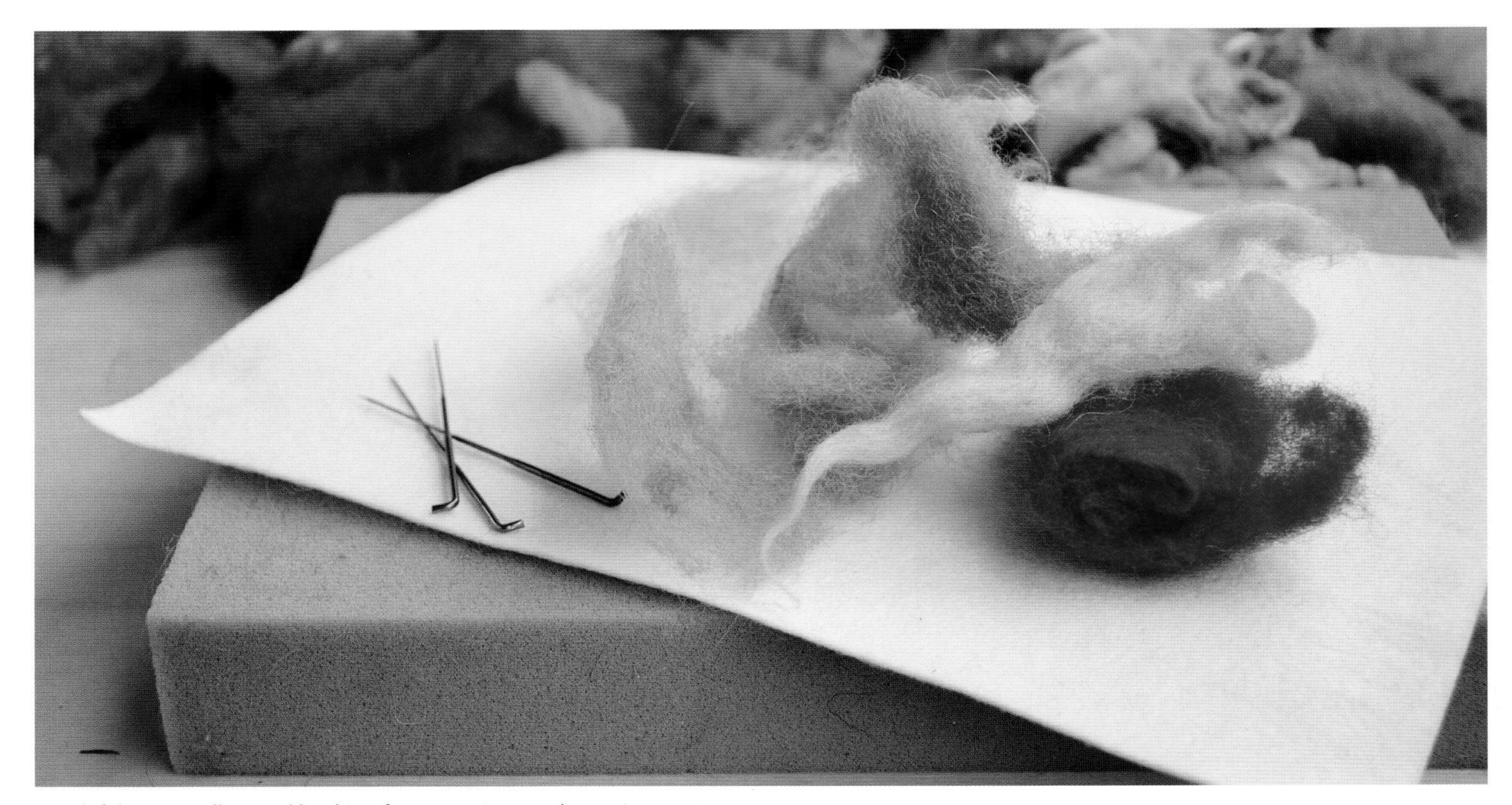

Wool, felting needles, and backing foam—you're good to go!

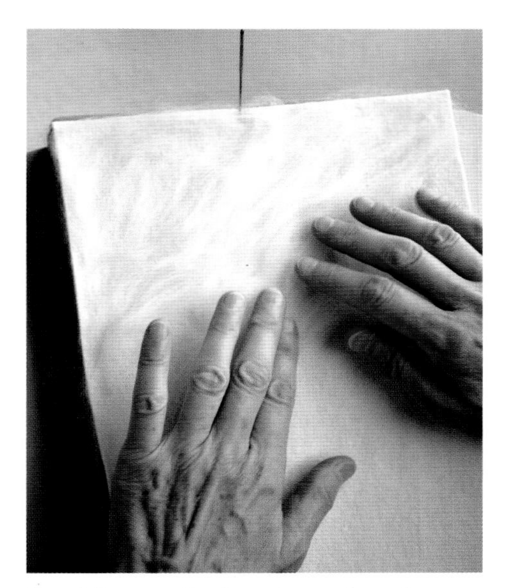

Carefully press the fine pieces of wool onto the backing foam.

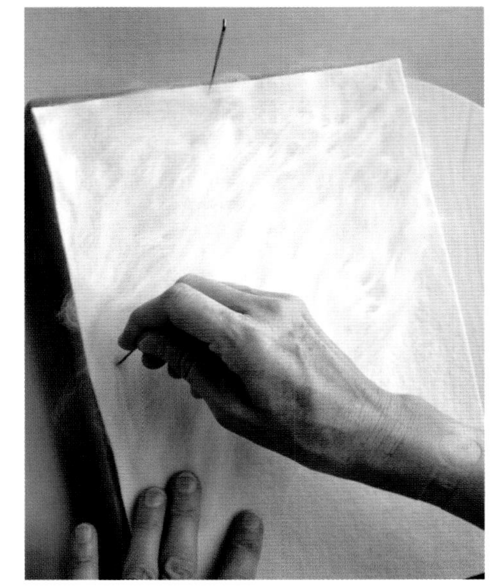

The wool is attached using the felting needle.

Now the next color layers are applied to the white background.

"line direction." Press the wool very gently now and again onto the felt backing with your hands. Once you have achieved your desired structure, take another felting needle in your hand and use this to stitch carefully through wool and backing. This way, the fine wool fibers become somewhat matted or felted together with the felt backing and will no longer come off. When you have prepared the background with the white wool, then the next step is "painting" with the wool.

When painting with wool, you can definitely work in a "breezy" or "airy" way, meaning that you shouldn't use any pieces of wool that are too dense. The more finely you apply the individual colors, the easier it is for the picture to create an effect. To design the sky, for example, you can use different shades of blue between which the white background shows through, or bring other colors into play, such as pink tones for a morning sky or orange tones for an evening sky. Press the individual color layers and dyed pieces down again carefully with your fingers and then gradually attach them with the felting needle. Don't attach them too tightly. To let the background show through slightly, it is enough to just attach the small pieces of wool in one

Delicate yellow, pink, and blue tones create the sky.

place. For denser motifs such as branches, on the other hand, you can attach them firmly to get the necessary density. To do this, use the felting needle to stitch through wool and

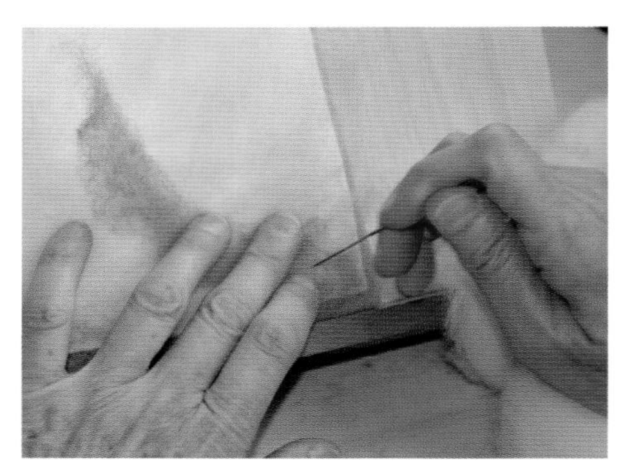

Use the felting needle to form brown wool into a branch.

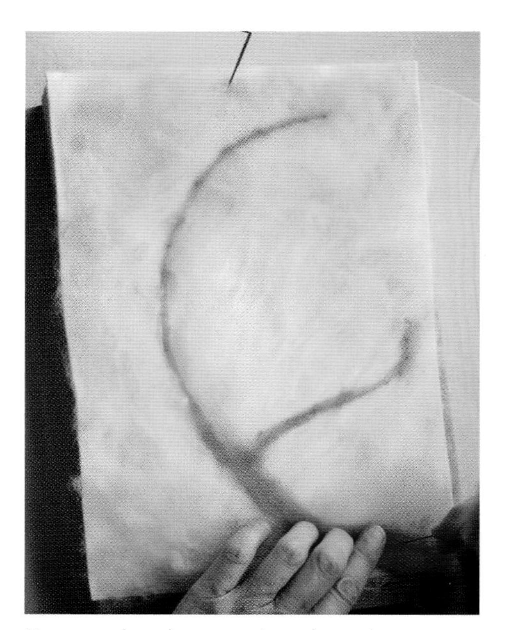

You can already recognize a branch against the airy background.

backing at an interval of a quarter inch or less. So that the wool becomes beautifully dense, you should work with several layers of this color shading on top of each other.

The design is created out of a wide range of colors.

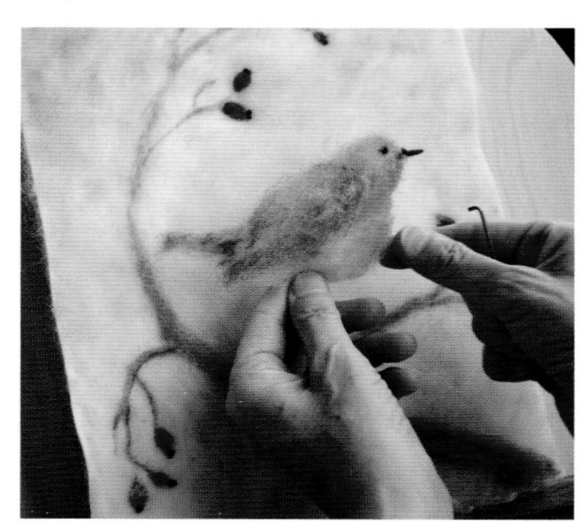

Smaller parts of the picture can be worked on separately.

If you want to insert smaller, compact picture parts, like the bird in our example, you can also prepare these on a separate, appropriately cut piece of felt and finally attach them when you finish the picture. This makes the work easier. Prepared sections can be attached to the background easily by felting the attached pieces with the felting needle. When you are satisfied with your work, you have to attach something on the edges so that the picture won't get detached from the felt backing, and then take it carefully off the foam pad. Wool pictures—presented in a frame—are a wonderful gift or a loving decoration for every child's room.

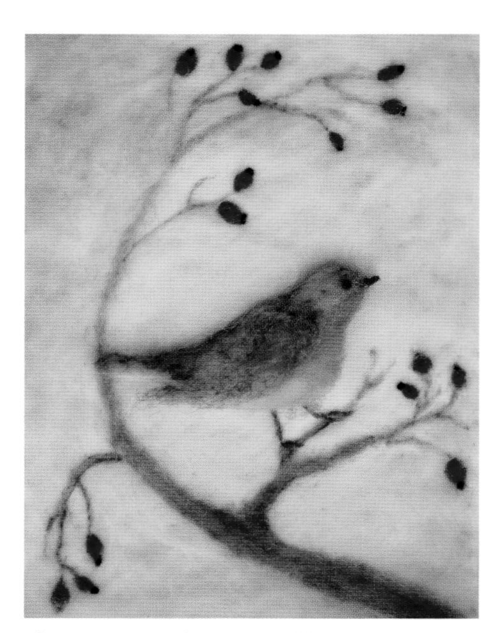

The imaginatively designed bird on a rose brier.

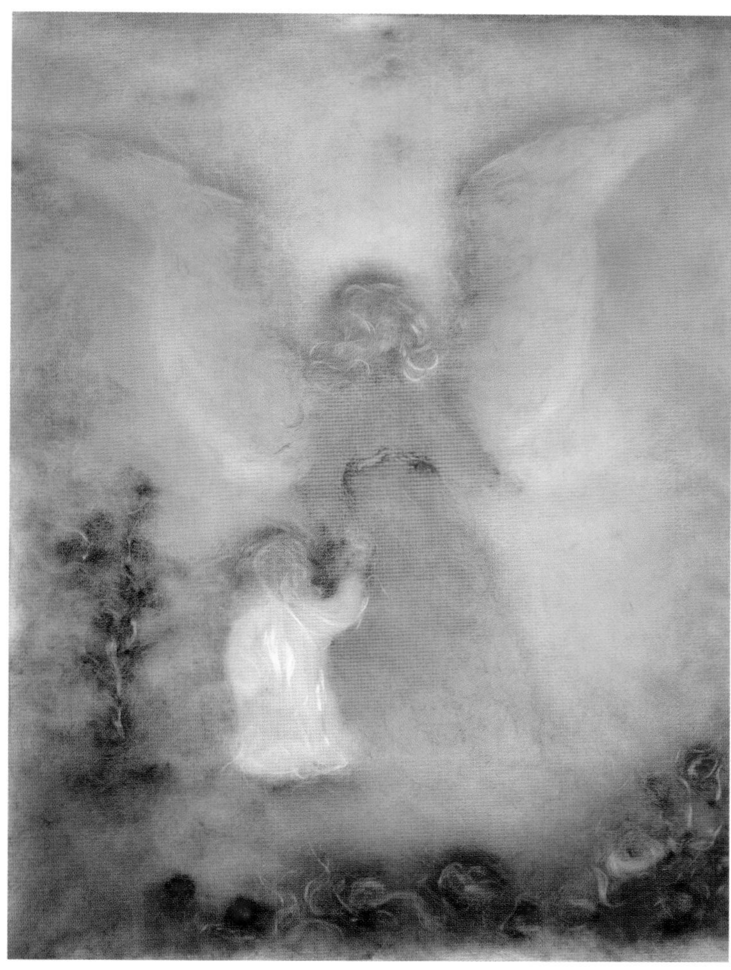
A guardian angel watches.

A luminous flower arrangement.

MAKING FELT FLOWERS

Felting is an ancient handicraft art, which is associated with many stories and legends. The oldest indications of the technique of felting come from Central Asia, but the use of felted clothing was already widespread among the ancient Greeks and Romans. It was customary for the Romans to give freed slaves a felt hat to wear, and this hat became a symbol of freedom. It is said that after the death of Nero, the Romans celebrated one and all, wearing felt hats on their heads in the streets, making the felt hat a political symbol.

When felting with wool, the structure of the material is destroyed by mechanical application, which makes the texture extremely compressed—this way, it is possible to make robust, waterproof garments. With the arrival of cotton to the West and then synthetic fibers, the technique of felting was more and more forgotten. In the 1980s, however, this way of treating wool experienced a big renaissance. Here we explain how to make small pieces of jewelry out of self-dyed wool; making larger pieces takes some practice. We will start with felting small flowers, to use for making necklaces.

Felt flowers in all the colors of the rainbow.

Wool, soap, water, and you're ready to go.

To make felt flowers you need:

- Unspun wool in various colors
- Soap
- Water
- A large piece of bubble wrap

Step by Step

To make a felt flower, use two hand-sized pieces of wool fleece, each of which you will work separately. Lay the bubble wrap on your work surface; on the one hand it serves to protect the work surface, but also provides resistance for the material you are working on. Theoretically, a washboard would also work for this. The soap should be as natural as possible, since you are working with your hands a lot, such as a good olive oil soap. Grate it small and dissolve in a little hot water. A teaspoonful of grated soap is enough to make a single flower.

Before starting the felting, divide the piece of wool fleece into four fine layers, and lay one atop the other crosswise. Thus you lay a fine piece of wool fleece on the wrap, then the next, with the direction of the fibers turned 90 degrees on top, and so on. If you don't do this, the grain of the fibers will run in just one direction in the material, making it harder to hook the individual fabric fibers together—which is the basic process of felting.

Now pour a little of the soapy water onto the fleecy "packet" of wool and then some hot water, just enough to dampen the wool well but so that the soapy solution doesn't drip off, but stays on the wool.

Now the actual felting work begins. Use both hands to rub the wool firmly over the bubble wrap in all directions, so that you get the shape you want. To make a felt rose, you need two half-oval pieces of felt of about the same size, and you should make sure that you don't stretch out the pieces too much. Whether you rub using outstretched fingers or prefer to use the balls of your hands is up to you. When working the wool, you will suddenly feel how the material is getting denser as the wool fibers laid atop each other catch into each other. The felt piece is finished when you have a compact, thin piece of felt that is the shape you wanted to create.

Pour the soapy solution carefully over the wool.

Rub the wool gently over the bubble wrap.

You've created the desired shape.

Use the water sparingly; you can always pour on more if needed.

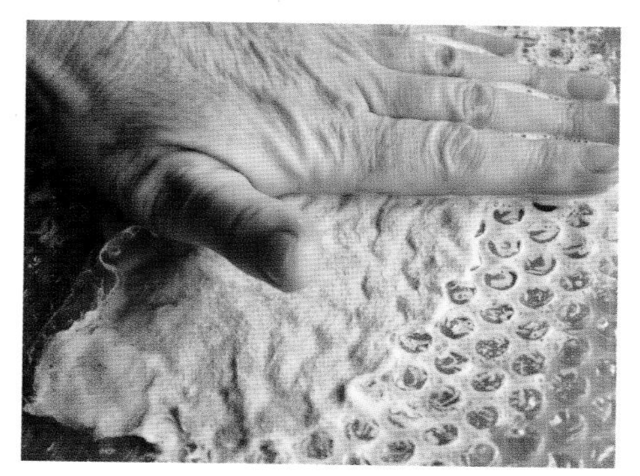

Use your fingers and the balls of your hands.

Rinse the finished felt pieces, thus, two half-ovals to make a rose, under running water until there is no soap left in the fabric. Then let them dry. As with all felting work, it also applies here that it doesn't matter if there are still fibers sticking up or there is no clean edge in places that will later be sewn or bound off. For the rose, we took care that the top edge of the petals is clean, while this isn't

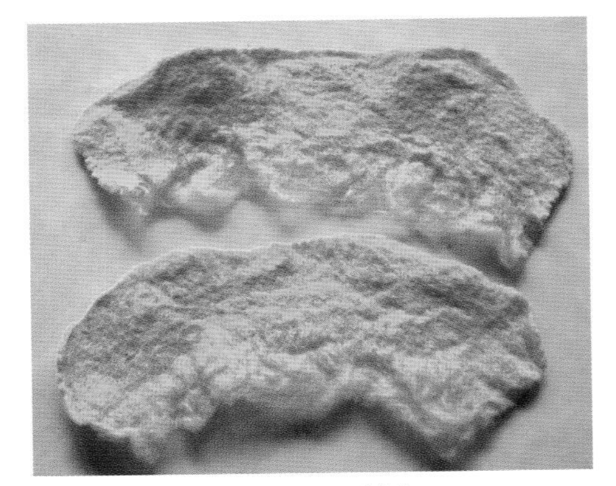

Wash out the two pieces of felt and let them dry.

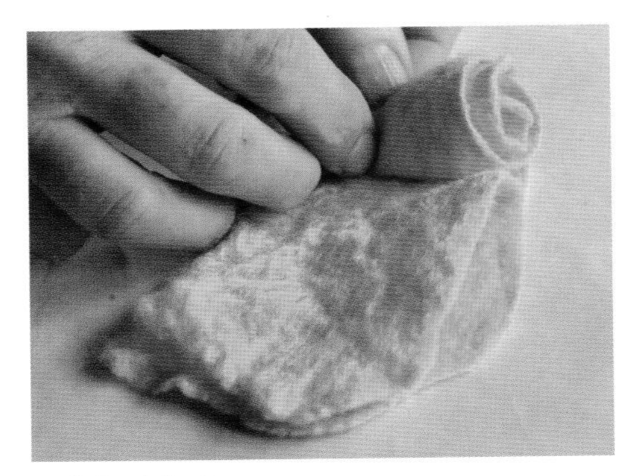

Roll the felt pieces up to make a flower.

Finally, sew the flower onto the "stem base."

necessary where the flower meets the stem. These places are also better for sewing because they can be shaped better than the firm processed edges.

Now lay one piece of felt atop the other and roll it up into a flower. Finally, sew it together so that it retains its shape. The flowers can be used for different decorations, can be worked up into jewelry, or help make a pretty gift wrapping.

Flowers used in pieces of jewelry—felted wool flowers with silk threads.

SPINNING

It is a special experience to spin wool you dyed yourself. Spinning is deeply embedded in many cultures and languages. In the Nordic mythological world, it was the three Norns who spin the threads of fate of human beings, and we know from fairy tales that the spinning wheel can be important; just think of Sleeping Beauty, who pricked her finger on a spindle. You no longer have to fear such injuries today, because you can't prick yourself on today's hand spindles.

When spinning, you can set the thickness and structure of the thread by your own preferences. Thus, the hoods, scarves, and headbands knitted from this wool become very special, unique pieces.

You should learn spinning under instruction, since it is barely possible to teach yourself this technique. However, there are also very precise instructions in the literature for spinning, which make it possible to at least take the first steps.

For spinning you need:

- A spinning wheel or a hand spindle
- A carder for combing the wool
- Unspun wool

Before the wool can be spun, make sure that it isn't matted. Wool can get matted from dyeing or removing remnants of plants and then you can't spin it. First you have to comb it, which is what carding means. You can do this with a simple hand-carder or a carding

Carding with a hand carder.

Blue and green wool are carded together with a carding machine.

Carefully remove the carded wool from the carding machine.

Twisting the thread with a Lazy Kate.

Spinning a single thread does not make a finished yarn, since the thread is plied in one direction, which would lead to it being weighted in one direction when it is knitted. Therefore a spun thread must be plied before further processing. To do this, two equally full spindles are wound off on a so-called "Lazy Kate," as shown in the pictures.

You get equally full spindles, in that the wool that is wound on the separate spindles is measured off before spinning. The wool is wound off from the Lazy Kate onto a spindle, and thus yields a yarn that isn't plied in one direction. When doing this process, you must take care that the spinning wheel is running in the other direction, as when spinning the individual spools.

machine. Carding also works for processing wool remnants, which are combined into one strand by carding and can then be spun. Moreover, you can card different colors together with each other. This makes it possible to spin different colored wool into one strand.

After carding, you can start spinning. You can spin both with a hand spindle and with a spinning wheel; using a spinning wheel allows faster and more accurate work. This isn't intended to be an exact guide to spinning—I would just like to present some possibilities for processing your self-dyed wool.

The spun wool.

Spinning the wool.

DYEING EASTER EGGS WITH PLANT COLORS

Dyeing Easter eggs with plants—this tradition is firmly embedded in Christianity, but it has also long been a tradition in other cultures. Decorated ostrich eggs that are several thousand years old have been found in southern Africa, and also have been found in the tombs of the ancient Egyptians, which are more than 5,000 years old. In the Christian tradition, painting and dyeing eggs is equally common, one of many customs that has made its way into Christianity from earlier times. Before the production of chemical colors and color additives, it was customary to dye the eggs with plants, just like textiles. In the peasant tradition, it is customary to use onion skins for dyeing, but Easter eggs can also be wonderfully decorated with other dye plants. Here we will give some examples.

Dyeing Easter eggs is always a joy, especially for children. Even if the little ones have to use a little patience to create the version with the flowers and

These eggs give a natural theme to your decor.

Easter may be on the way!

Chamomile, herbs, bloodwood, and madder root—everything for a great dyeing experience!

Fresh herbs for decorating eggs.

What else do you need?–old stockings, scissors, a sponge, and some twine.

leaves, the experience of success is worth the effort!

To dye the eggs, in this book we used chamomile, bloodwood, madder, turmeric, onion skins, and matcha tea, but an innovative and experimental dyer should realize that they can try using all the dyestuffs also used for textiles! The eggs can be dyed one color with the selected dyestuff, but we will also demonstrate a very special dyeing technique, which is to apply plant patterns to the Easter eggs.

Eggs with Plant Patterns

To prepare for dyeing, look in the fields and meadows for herbs, leaves, and flowers that you would like to "portray" on the Easter eggs. Leaves and flowers with a firm structure work especially well. Since applying the plant pieces to the eggs takes some time, put the plants in water beforehand so that they don't get wilted.

Both white and brown eggs can be dyed well, except that you won't be able to dye brown eggs with the green colors from matte tea. Before you start dyeing, set everything that you need out ready: the eggs, the plants, and the dyestuffs as well as old nylon stockings and pantyhose, a scissors, a sponge, and enough twine to tie things up.

Some dyestuffs must be soaked overnight, just like for dyeing textiles. If you are using enameled pots, you should know that the dyestuffs will also dye the pot heavily. You should therefore use separate pots for dyeing Easter eggs, and not those you will use again to prepare food. This applies in general for dyeing textiles—you should store your equipment for the dyeing studio separately, and not use it again for cooking! The dyestuffs that have to be soaked in a pot of water overnight include onion skins, madder root, and bloodwood, although bloodwood also works without soaking. This is not the case for onion skins and madder root!

Soak the onion skins overnight.

An egg dampened with the sponge is decorated with herbs.

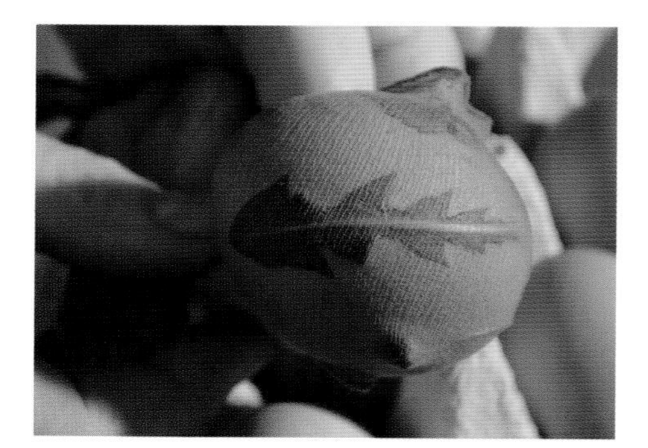

The egg with the herbs on it is tied into a stocking.

The eggs packed in stockings.

Step by Step

Prepare the raw eggs for the dyeing process. First, if necessary, clean every egg of dirt and feathers; you can use the sponge for this. Use it to dampen the egg again after cleaning, so that the herbs and flowers adhere to the egg shell better. Each egg can be individually designed by placing the leaves and flowers on the egg and pressing lightly. Next carefully put the egg in a piece of stocking and then tie it up with twine. The nylon stocking should hold the egg firmly so that the leaves don't slip off. Now prepare all the eggs that you have chosen to dye this way.

Boil the soaked **onion skins** for about half an hour. Here you should note that water evaporates off during the boiling and you should put the skins in enough water so enough remains in the pot to cover the eggs after the first boiling step. For one dyeing, two or three handfuls of yellow onion skins should be enough. These dye the eggs golden yellow to brown, depending on the amount of skins. Before you put the eggs in the pot, the onion water should be cooled down again; during this time you should carefully prick the eggs with an egg pricker or a pin, which prevents the shell from cracking open. Boil the eggs in the onion water, between seven and ten minutes as desired. Then take the eggs out of the water, rinse them with cold water, and take them out of the stocking.

Put the wrapped eggs in the pot.

Eggs colored with onion skins.

Grate some cochineal insects in a coffee mill.

Small amounts, but great dyeing results.

Cochineal dyed eggs

When dyeing with **madder**, soak a handful of madder root overnight, and the next day boil the dye bath for about an hour, with the addition of a teaspoon of vinegar. Then boil the eggs in the egg dye, and the result is brick red to rust red.

If you want to have pink and rose-colored eggs, use a tablespoon **cochineal** for dyeing them. The dried cochineal scale insects are ground in a coffee grinder or finely crushed in a mortar. This very small amount of cochineal produces wonderful intense color results. To do this, boil the powder for five to ten minutes and then let the water cool and add the eggs to the lukewarm liquid. Bring it back to a boil and boil the eggs again for at least seven minutes. After the eggs are cooked, rinsed in cold water, and unpacked, let them cool and then grease them with some butter or olive oil. This makes the eggs shine and gives radiant colors .

You can dye Easter eggs a dark purple with **bloodwood**. Boil a small handful bloodwood in a pot with some water for half an hour. Then let the dye bath cool, add the eggs, and bring it to a boil again. Eggs dyed with bloodwood show a very intense color result and the plants tied in the stocking show to wonderful advantage.

For a second version, take the dampened egg, put it in the stocking, and then spread a little bloodwood on top. Now add any other dyestuffs you want, such as a few pieces of madder root, chamomile, or other dyeing

Eggs dyed with bloodwood.

Put the egg in the stocking and sprinkle with bloodwood.

The second method for bloodwood eggs.

Little effort, great effect: lemon juice for painting the eggs.

Paint the lemon juice on with the brush.

The finished decorated eggs.

flowers. Tie the stocking tightly so that the dyestuffs remain stuck to the egg. Boil the eggs prepared this way in water for at least seven minutes, and then remove them and rinse in cold water. When your remove the stocking you can see the colorful patterns, and each egg is unique. This version can be combined with the application of leaves and flowers!

A third way of dyeing with bloodwood is to treat the bloodwood-dyed eggs with lemon juice. To do this, squeeze a lemon into a small bowl. Use a toothpick or a very fine brush to draw on the dark egg with the lemon juice, as if using watercolors. Wherever the lemon juice is applied, the color disappears again, so you can write on your eggs individually and decorate them with different designs.

Dyeing Easter eggs with **matcha tea** creates something very special. The green color doesn't like to develop on brown eggs, so here you should use white eggs that are cooked before the dyeing process. Then boil a handful of matcha tea for an hour in the dyeing pot and add the eggs to the lukewarm liquid. Leave them there until they turn green, which can definitely take an entire night.

Eggs dyed with matcha tea.

Onion skins, cochineal, and bloodwood—all kinds of splendor.

BIBLIOGRAPHY

Bächi-Nussbaumer, Erna. *How to Dye with Plants*. Bern/Stuttgart: Haupt, 1980.

Berger, Dorit. *Dyeing with Natural Colors: Dye Plants, Recipes, Possible Applications*. Stuttgart: Ulmer Verlag, 1998.

Bogusch, Graf; Schnalke, ed. *On Life and Death: Contributions to the Discussion about the Exhibition "Body Worlds."* Steinkopff Verlag, 2003.

Clausen, Anke-Usche, and Martin Riedel. *Systematic Workbook, Volume IV: Designing Colors Creatively with Their Associated Materials*. Stuttgart: Mellinger Verlag, 1981.

Crook, Jackie. *Dyeing Naturally*. Stuttgart/Bern/Vienna: Haupt, 2008.

Finlay, Victoria. *The Secret of Colors: A Cultural History*. Berlin: Ullstein Buch Verlag, 2005.

Fischer, Dorothea. *Dye Wool and Silk with Natural Materials*. Aarau: AT Verlag, 1999.

———. *Natural Colors on Wool and Silk*. Norderstedt: Books on Demand GmbH, 2006.

Gromer, Karina. With contributions by Regina Hofmann-de Keijzer and Helga Rösel-Mautendorfer. *Prehistoric Textile Art in Central Europe—History of Handicrafts and Clothing before the Romans*. Vienna: Publications of the Prehistoric Department of the Natural History Museum of Vienna, the Natural History Museum Press, 2010.

Hirsch, Siegrid, and Felix Grünberger. *The Herbs in My Garden*. Unterweitersdorf: Freya, 1999.

Hofmann. "The Colors and Dyeing Techniques of Prehistoric Textiles from Salzburg Hallstatt." In Gromer, K., et al., *Textiles from Hallstatt: Weaving Culture from the Bronze Age and Iron Age Salt Mines*. Budapest: Archaeo-lingua, 2013.

Karl, Andrea, and Maria Karl. *Dyeing and Felting*. Graz: Leopold Stocker Verlag, 1996.

Leitner, Christina. *Paper Textiles*. Bern/Stuttgart/Vienna: Haupt, 2005.

Nencki, Lydie. *The Art of Dyeing with Natural Substances*. Bern/Stuttgart: Haupt, 1984.

Nübling, Eugen. "Ulm's Wine Trade in the Middle Ages, Ulm 1893," www.ulm.de/sixcms/media.php/29/1_4_M4.pdf.

Pastoureau, Michel. *Blue: The History of a Color*. Berlin: Verlag Klaus Wagenbach, 2013.

Ploss, Emil Ernst. *A Book of Old Colors: Technology of Textile Colors in the Middle Ages*. Gräfelfng: Verlag Moss & Partner KG, 1989.

Prinz, Eberhard. *Dye Plants*. Stuttgart: E. Schweizerbart'sche Verlag, 2009.

Schneider, Gudrun. *Dyeing with Natural Colors*. Ravensburg: Otto Maier Verlag, 1979.

Schweppe, Helmut. *Handbook of Natural Dyes*. Landsberg/Lech: Ecomed Publishing Association, 1993.

Sherf, Gertrud. *Plant Secrets from Olden Times*. Munich: BLV, 2004.

Simonis, Werner Christian. *Wool and Silk*. Stuttgart: Verlag Freies Geisteswesen GmbH, 1977.

Weinmayr, Elmar. *The Rainbow Color Thief: Dyeing with Plants in the Yoshioka Workshop*. Kyoto: Shikosha Press, 2011.

ABOUT THE AUTHORS

Franziska Ebner lives in Salzburg, Austria, and has been dyeing with plants for about 30 years. She makes the dyed textiles into jewelry and decorative accessories, which she sells at artisan markets and exhibitions. Learn more about her at www.franziskaebner.com.

Romana Hasenöhrl is a freelance author and lives in Salzburg, Austria. Learn more about her at www.romanahasenoehrl.at.